D0251431

The Ape in the Tree

Alan Walker   Pat Shipman

The APE in the TREE

An Intellectual & Natural History of *Proconsul*

The BELKNAP PRESS of
HARVARD UNIVERSITY PRESS
Cambridge, Massachusetts
London, England
2005

Printed in the United States of America

Design by Jill Breitbarth

Title page art by John Gurche

Library of Congress Cataloging-in-Publication Data
Walker, Alan, 1938–
The ape in the tree : an intellectual and natural history of Proconsul /
Alan Walker and Pat Shipman.
p. cm.
Includes bibliographical references (p.).
ISBN 0-674-01675-0 (alk. paper)
1. Proconsul. 2. Apes, Fossil. 3. Paleontology—Miocene.
I. Shipman, Pat, 1949– II. Title.
QE882.P7W35 2005
569′.88—dc22 2004057405

To the memory of John Russell Napier (1917–1987)

Contents

Author's Note

I have tried throughout this book to give credit to the many friends and collaborators who have played a part in the *Proconsul* story as I see it, but I make no claim to have covered all the research or mentioned all the researchers whose work bears on our understanding of this ape in our tree. This book is intended to serve as a small taste, flavorful enough to attract readers and provoke their curiosity yet small enough to leave them with an appetite for more.

In the interests of telling the story well, I have not stopped at every point to thank the many foundations that have generously provided support for my work discussed here. These have been the National Geographic Society, the National Science Foundation, and the Wenner-Gren Foundation.

Many people at several institutions have also provided invaluable assistance with historical information, notably: Harriet Ritvo of MIT; The Natural History Museum in London and its archivist, Susan Snell; the National Museums of Kenya and its incomparable staff; the Bodleian Library of Oxford University and its archivist, Colin Harris; the Keeper of the Archives at Cambridge University Library, Jacqueline Cox; and the interlibrary loan staff of the Pennsylvania State University library. People who kindly and carefully read portions of the manuscript or checked information include David Begun, Chris Dean, Wendy Dirks, Jay Kelley, Russ Tuttle, Holly Smith, Tanya Smith, and Fred Spoor.

I also want to thank the Leakey family, the Hominid Gang, the people of Rusinga and Mfangano Islands, and everyone who participated in the various field expeditions and analyses for their help and friendship. Editor Michael Fisher and agent Ralph Vicinanza have been invariably helpful and enthusiastic.

Many of the places where important discoveries have been made are in Africa and have names that will be unfamiliar to some readers. There is a guide to pronunciation at the back of the text.

Though the stories in this book are mine, and are told in my voice, nearly all of the writing has been done by my wife, Pat Shipman, for the simple reason that she is a better writer than I am.

Alan Walker
State College, Pennsylvania

The Ape in the Tree

Prologue

He caused a sensation at the Folies Bergère in Paris. Elegantly attired in a custom-made tuxedo, he strode onto the stage and tipped his top hat to the crowd; they applauded wildly. Nearly all attending the show had heard of his highly successful tours of America and other cities in Europe.

His act was deceptively simple. He presented himself, played the piano, rode a bicycle, and then, as if tired by his exertions, sat down to eat a quick meal with a glass of wine. Afterward, he smoked a cigarette. As a finale, he stood on his head, undressed to his pantaloons, and somersaulted into bed.

He was a chimpanzee named Consul and it was 1903.

This was more than a simple animal act. Dogs that jumped through hoops, ponies that trotted in formation, even the occasional lion or tiger were commonplace in vaudeville shows and circuses at the time. What Consul offered was something more, something that led to a questioning of the fundamental principles of identity. Consul fascinated his audiences because he revealed the biological closeness of humans and apes. He "aped" humans, as the contemporary report of his act in *La Nature* remarked. He behaved like a civilized, even well-to-do, human though he appeared thoroughly apelike, for even the custom-made tuxedo could not disguise his long arms, hairy

Consul the performing chimpanzee, shown here at the Folies Bergère in Paris in 1903. His act fascinated audiences because it blurred the line between human and ape. (From *La Nature* 1591 [1903]: 416.)

skin, small braincase, and elongated toes. The blurring of such important and distinct categories as "ape" and "human" gave the audience a delicious frisson of danger, delight, and confusion. Was he a chimpanzee? Yes, patently so. For all his élan, there was no mistaking this individual for a human. He was mute, hairy, misproportioned for a human. But did he dress, walk, eat, cycle, and smoke like a human? Yes he did. Had he been able to tap-dance and sing a song, the illusion might have been complete.

So appealing was Consul's act that he was but one of a succession of performing chimpanzees named Consul. He—or they—were famous far beyond what might have been expected. The name Con-

Another Consul lived and performed in Belle Vue Zoo, Manchester, England. His death in 1894 was honored on a specially published broadsheet. (From Chetham's Library, Manchester.)

sul became eponymous for zoo or performing chimpanzees, even as Jumbo, a bull elephant originally exhibited at the London Zoo and then sold to the Barnum & Bailey Circus, came to mean "elephant"—and, eventually, anything huge.

There was even an obituary printed for an early Consul of the Belle Vue Zoological Gardens in Manchester, England, who died on November 24, 1894. At the top is a photograph of Consul, wearing a natty striped hat and matching blazer, like any smart British holidaymaker. He holds a pipe and sits in a rattan chair pulled up to a table bearing an empty glass and a bottle. Beneath the photo is a poem, written for the occasion by Ben Brierley.

"Hadst thou a soul?" I've pondered o'er thy fate
Full many a time. . . . Thou hadst ways
In many things like ours. Then who says
Thou'rt not immortal? . . .
'Tis God alone knows where the "Missing Link"
Is hidden from our sight; but, on the brink

Of that Eternal line where we must part
For ever, sundering heart from heart,
The truth shall be revealed . . .

It is doggerel, but it shows the widespread confusion about the exact distinction between *human* and *ape.*

The name Consul was so potent a symbol of the close resemblance of humans and apes that the British paleontologist Arthur Tindell Hopwood borrowed it in 1933 when he needed to create a new name for a fossil ape that he thought was ancestral to chimpanzees. He called the specimens *Proconsul africanus,* meaning "the African ancestor of Consul." Why did Hopwood do this? As a child born in 1897, Hopwood may have been impressed by one of the Consuls that toured Europe in the early part of the twentieth century. Perhaps he had only heard stories of performing chimpanzees called Consul, as I have; I do not know. He was probably being whimsical, yet there was an important message in his choice. By formally naming these fossils after Consul, he made a lasting reference that had great meaning to his peers.

Like every poor, performing chimpanzee, Consul reminds us of Darwin's awe-inspiring and wonderful theory of evolution—and of the descent of humans from apelike ancestors. From the written descriptions and photographs of Consul, I believe that he was perceived (by nineteenth- and early twentieth-century standards) as more human superficially than many of the indigenous peoples in far-flung corners of the European colonies or even than many of the desperately poor of Europe.

By blurring the behavioral distinction between humans and apes, Consul raised real and awkward questions. Did the nature of humanity lie in anatomy or training? Did breeding always tell? Then how could an ape be taught to walk and dress and behave like an aristocrat? In 1913, when Consul would still have been vividly remembered, George Bernard Shaw wrote *Pygmalion,* a play in which a lowly Cockney flower seller is transformed into a beautiful young lady of society. The message of the play is that superficial things, like manners and dress, truly constitute aristocracy. Aristocracy and superiority lie not in an innate, inborn quality but in training and habit. That Consul lived offstage not in a dismal, barred cage but in a suite of hotel rooms with his manager—the Professor Henry Higgins of the ape world—intensified the sense that a barrier once thought inviolable had been crossed.

For some time, my colleague Morris Goodman has been reinvigorating the debate over the distinction between chimpanzees and humans with molecular evidence. He has suggested that chimpanzees are so like humans, molecularly, that they ought to join us in the genus *Homo.* Many people find this idea profoundly disturbing.

From its outset, then, the story of the fossil *Proconsul* has been an odd and quirky one, full of symbolism, jokes, and surprises. To understand *Proconsul* I have had to grapple with what it is to be an ape, and a human. Studying *Proconsul* is attempting to unravel the secrets of our past—*ours,* the one we share with all the apes: chimpanzees, bonobos, gorillas, orangutans, siamangs, and gibbons. *Proconsul* is not just a human ancestor but also an ape one, the last common ancestor to whom we—humans and apes alike—all trace our past.

It has been my good fortune to be involved in a large part of the story, although its beginning with Arthur Tindell Hopwood preceded my birth. At the beginning of the twenty-first century, it seems fitting to look back at how much, and how little, we have learned about this extraordinary creature *Proconsul.* Hopwood had only a jaw and some teeth, but this first find was soon surpassed by others. There was a wonderful skull, found in 1948, and a nearly complete arm and some foot bones unearthed two years later. There were jaws here, the odd vertebra there, a leg bone at another place. Then, working in the 1980s on Rusinga and Mfangano Islands in Lake Victoria, my colleagues and I stumbled upon hundreds of fossils representing virtually every part of the body of *Proconsul.* We recovered many different individuals, including partial skeletons of infants, juveniles,

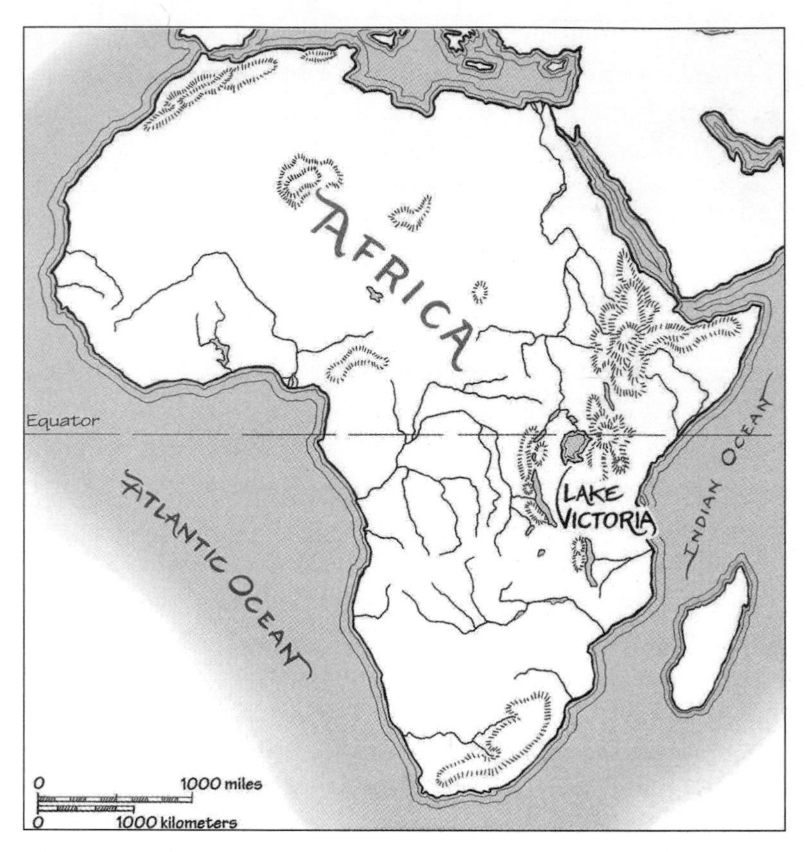

Africa. (Map by Jeff Mathison.)

Facing page: Top: East Africa, where all *Proconsul* fossils have been found. (Map by Jeff Mathison.) Bottom: My team and I excavated hundreds of fossils of *Proconsul* from sites on two small Kenyan islands in Lake Victoria: Rusinga and Mfangano. (Map by Jeff Mathison.)

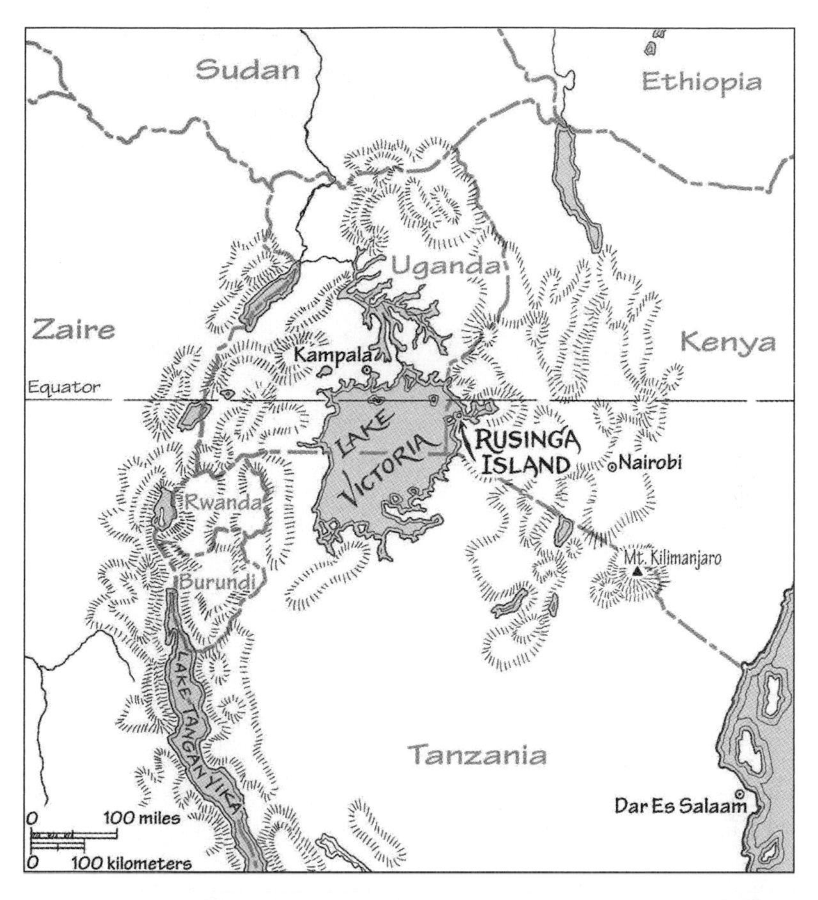
Sudan
Ethiopia
Uganda
Zaire
Kenya
Kampala
Equator
LAKE VICTORIA
RUSINGA ISLAND
Nairobi
Rwanda
Burundi
Mt. Kilimanjaro
LAKE TANGANYIKA
Tanzania
Dar Es Salaam
0 100 miles
0 100 kilometers

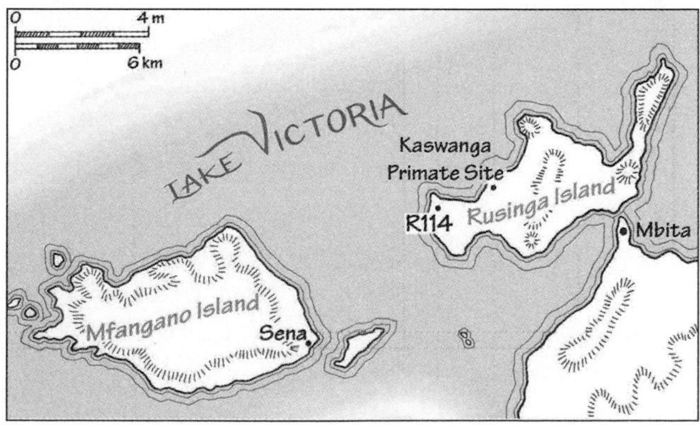
0 4 m
0 6 km
LAKE VICTORIA
Kaswanga
Primate Site
R114
Rusinga Island
Mbita
Mfangano Island
Sena

and adults, both male and female. We excavated parts of skulls and teeth and vertebrae, limbs, hands, and feet of *Proconsuls* of all sizes. Many of our finds came from a place we called the Kaswanga Primate Site on Rusinga, where there was literally an embarrassment of riches.

All *Proconsul* specimens have come from sites either in Kenya or in nearby Uganda, and they are all associated with deposits created by ancient volcanoes with mellifluous African names: Tinderet, Moroto, Napak, and Kisingiri. Several different species have been recognized within the genus *Proconsul*, in addition to *P. africanus*, and they varied from the size of a large monkey to the size of a gorilla. I can tell when they lived with considerable precision—all specimens are dated to between 21 and 14 million years—and I can outline at least some of the geographic range they once inhabited.

I know how they grew from infancy to adulthood.

I also know an amazing amount about their habitats and habits: the things they ate and the things that ate them, thanks to my own work and that of my colleagues. Some of the fossil *Proconsul* sites preserve fossilized leaves, fruits, caterpillars, grasshoppers, and other wonders, some of which must have figured (when fresh) in *Proconsul*'s diet. It is a case of turn and turn about, though, for some of the *Proconsul* fossils have been chewed by sharp teeth and are undoubtedly the remains of a predator's dinner, as are some of the other creatures found with *Proconsul*.

Most of all, I know that there was an evolutionary blossoming, what is technically called an adaptive radiation, of these strange and yet familiar Miocene apes. An adaptive radiation occurs when a single stock of animals—in this case, a primitive apelike creature—gives rise to a wide range of descendant species (this is the "radiation") that differ in various ways because they have adapted to fill many different ecological niches (this is the "adaptive" part). A case often used to illustrate the meaning of the term is that of Darwin's finches. Seemingly, when the first finch species reached the Galápagos Islands, so few birds had managed to reach them—the islands are about 600 miles from the mainland—that many ecological niches or ways of life filled by various types of birds in South America were vacant, leaving potential resources unexploited. Over time, the finches evolved, adapted, and diversified, their beak shapes and proportions changing until the finches filled fourteen different niches. One finch became a sort of functional equivalent of a woodpecker, another a small seed eater, yet others became ground dwellers while others still remained perched in trees; some eat insects while others are strictly vegetarian.

How does this concept apply to Miocene hominoids? We have to look to African forest monkeys to see a situation among higher primates that resembles what was going on during the early Miocene. Today, there are at least twenty species in the genus *Cercopithecus,* including the bearded De Brazza's monkey, the beautifully marked Diana monkey, the spot-nose guenon, the blue monkey, and many others. In some particularly rich forested areas, numerous species of cercopithecine monkeys can and do coexist, along with olive, red, or black and white colobus monkeys. If you look in the deep forest, you might find drills or mandrills, with their brightly colored faces and scarlet rumps, as part of the primate community. These varied Old World monkey species sometimes feed and travel in mixed species groups, recognize and react to each other's alarm calls, and seem to benefit more than they are hurt by sharing their habitat with the other species. Behavioral ecologists today know a good deal about how these varied monkeys partition up the forest and its resources, which allows them to coexist without crippling interspecies competition.

As we look at the Miocene sites in Africa, we see a similar pattern except that the fossil primates are early apes, not monkeys. There were many hominoid species, undoubtedly more than we know, that seem to have been but minor variants on an apelike, primitive evolutionary theme. And somewhere within this adaptive radiation lie the origins of some very important branches on the primate family tree, including the common ape-human stem that eventually split into separate lineages leading to each of the modern apes and to ourselves.

Proconsul is the best candidate we know of for the last common ancestor of both modern apes and humans, the very trunk of that forking, branching tree that is so fascinating because it is the record of our own origins.

Proconsul is neither ape nor human, but something else: something ancestral, extinct, and fascinating. In more ways than one, *Proconsul* is truly the ape in the tree, the ape in *our* tree.

1

Luck and Unluck

Dayrell Botry Pigott was an unlucky man.

Fossil-finding has always been a somewhat risky occupation, for fossils are often found in remote badland regions, where erosion by wind and water concentrate the petrified bones on the surface in lag deposits. If you are hunting for fossils in Africa, then, you expect to have to watch out for snakes, scorpions, and sunstroke, bad roads and worse drivers, and an uncertain supply of provisions. To find human or primate fossils, you usually have to drive halfway to the ends of the earth to get to where there might be fossils. Then sometimes you meet a local inhabitant who is convinced you are mining for gold, diamonds, or some other priceless treasure on his land, and he doesn't like the idea. The truth—that you are looking for bones that have turned hard like rocks, that are millions of years old and will tell you about your long-long-distant ancestors—sounds too absurd to be believed in many instances. There are also the odd tropical and not-so-tropical diseases to watch out for and insidious parasites that lurk in earth, food, and water, waiting to infect unsuspecting humans. Perhaps most dangerous of all is that scourge of the modern world, political turmoil, but car accidents run a close second.

What you don't expect is to be eaten by a crocodile.

Pigott probably didn't expect it either. In 1909, he was just 30 years old and the assistant district commissioner of the Kavirondo region in western Kenya Colony (as it was then). He was sent out to the town of Karungu, to check out some fossils that had been reported to a British official, C. W. Hobley.

Hobley was something of a polymath. Born in Chilvers Coton in Warwickshire, England, Hobley had had a strictly technical education in engineering at the Mason Science College (now the University of Birmingham), which specifically proscribed literary and theological subjects. As might be expected, Hobley was highly competent in many matters important to building a colony, such as geology, water supplies, road building, surveying, and establishing close and cooperative relations with leaders of the local tribes. Despite his focused education, however, Hobley had eclectic interests. He took full advantage of the novel opportunities that surrounded him in the East African colonies, writing on such diverse subjects as snakes, rhinoceroses, pottos, crocodiles, elephants, serval cats, wildlife conservation, tribal methods of hunting and trapping, tribal beliefs and magic, human origins, ethnology of the Wakamba, El Dorobo, Nandi, Kikuyu, Turkana, Suk, and Zanzibaris, archaeology, and, among other geological subjects, volcanoes, soil erosion, earthquakes, and ancient environments. Little wonder, then, that when the jaw of a deinothere, an extinct relative of the elephant, and other fossils turned up in Kavirondo someone sent word to Hobley.

At the time of the discovery, Hobley was no longer the provincial commissioner of Kavirondo, the region now known as Nyanza Province, where the fossils had been found. Hobley was sub-commissioner of Ukamba Province, stationed in Nairobi, and he later wrote a famous monograph about the Wakamba people for whom the province was named. Even though he had left Kavirondo before 1909, Hobley's friendships with the white settlers in that region continued. When Hobley learned of the fossils, he in turn asked Pigott, who was still working in Kavirondo, to stop by and examine the specimens and the site when he was next in the area.

I have been able to discover a little of D. B. Pigott's family history.

He came from a line of country vicars with impressive names; his father was the Reverend Eversfield Botry Pigott and his grandfather was the Reverend Shreeve Botry. D. B. Pigott went to St. John's School in Leatherhead, England, originally founded to educate the sons of clergymen as a charity but which, by the time D. B. enrolled, was attracting fee-paying students and had all the characteristics of some of the more famous British public schools. After St. John's, Dayrell went to the elite Magdalene College of Cambridge University for a B.A., graduating in 1901. In 1902 he was called to the bar and spent two years as a solicitor before becoming the private secretary to the administrator of Salisbury, Rhodesia: substantial responsibility for a young man. In 1907, he joined the Colonial Office to serve as an assistant district commissioner in Kavirondo. He'd been in that post for nearly five years before the crocodile got him on February 28, 1911.

D. B. Pigott was one of those young men of good family who had been educated in a prestigious secondary school before attending one of the superior universities, in his case Cambridge. In the nineteenth and early twentieth centuries, dozens of young men of such background were sent out at tender ages to rule the British Empire with astonishingly little training. It was generally believed that a public school education, a few years at university, a good moral character, and some ability in sport were entirely sufficient qualifications for administering enormous tracts of land with huge populations virtually anywhere in the British Empire. What amazes me is that such a cockamamie premise worked so well for so many years, not that it eventually came to grief when the former colonies demanded their dignity and independence.

The fossil deinothere jaw that started everything for Pigott was one of several important fossils of the Miocene age, a geological epoch that lasted from 23.5 to 5 million years ago, discovered by G. R. Chesnaye somewhere along the eastern shore of Lake Victoria. Pigott duly collected the fossils and sent them off to the British Museum (Natural History)—then colloquially known as the BM and now called The Natural History Museum—in London, where they captured the imagination of another man, Dr. Felix Oswald. Oswald persuaded Arthur Smith Woodward, the keeper of palaeontology at the British Museum, to help him raise private funds to go to Kenya and look for more fossils. By November 1911, this fund amounted to the grand sum of £270 and Oswald set off.

But in the meantime, Pigott had met with disaster. On February 28, 1911, while traveling by raft on Lake Victoria, Pigott had encountered a truculent hippopotamus. This is not an uncommon occurrence. It happened to me fairly often when my team and I were working on Mfangano Island in Lake Victoria and I had to ferry supplies back and forth from the mainland. Unfortunately, hippos sometimes feel boats are rivals. I wonder if they are simply unable to distinguish between a rival hippo and any large object that moves through the water. In any case, hippos sometimes react as if a boat were a strange hippo invading their territory or threatening their young. Their instinct is to attack by biting the interloper or by coming up underneath it, perhaps to attack its belly. When successfully carried out, these assaults pretty well negate the seaworthiness of a vessel that is usually heavily loaded with gear, supplies, and men, many of whom cannot swim. Hippo attacks are truly dangerous and life-threatening to those in boats. I always keep a nervous eye out for hippos when I am in a small boat in Africa, though my only counterstrategy is avoidance.

Apparently Pigott had the same worries but took a more aggressive stance than I ever did; he shot the hippo. Shooting can be perfectly effective if you kill the hippo instantly—and if you don't mind killing such a magnificent creature, which I would. If you merely wound the hippo, you enrage it and encourage the hippo to attack your boat all the more ferociously, which is apparently what happened to Pigott. Probably Pigott could swim, so drowning wasn't his immediate problem. His problem was the opportunists attracted by the commotion, at least one of which was a Nile crocodile.

In the days before hunting had diminished the populations of crocodiles, these creatures were very large. Hobley records that the duke of Mecklenburg shot a croc that was 21 feet, 6 inches long long in Lake Victoria and Hobley himself killed a monster 16 feet, 6 inches long, which nearly killed him instead. Even a small 5- or 6-foot-long crocodile is a formidable opponent that can kill a human without much difficulty. A crocodile's favorite technique is to grab someone firmly by an arm or leg and then keep its struggling prey under water until it drowns. So Pigott, incautious enemy of the hippo, became the lunch of the crocodile. In London, *The Times* reported his death "by drowning" and commented with admirable understatement, "His body has not yet been found." I suspect it never was found, at least not in one piece.

Pigott's primary tributes (from those outside his family) were the naming of a fossil crocodile in his honor, *Crocodylus pigotti,* and the persistence of an anecdote about the assistant district commissioner of Kavirondo who found fossils but was eaten by a crocodile. The sad thing is, in most written versions of the story I have seen, his name is given as *Digby* Pigott, presumably a confusion between his initials, D. B., and the upper-crust British name.

With Pigott went all knowledge of the exact location of the fossil deposits from which he had collected the fossils in Kavirondo. When Oswald appeared in Kenya Colony in November 1911, he had only vague, general information to go on.

Fortuitously, Oswald happened to encounter Chesnaye, who had found the fossils in the first place, on his way upcountry from Nairobi. Chesnaye could and did tell Oswald precisely where he had collected the original fossils, which was presumably the same spot that Pigott had found. Chesnaye also told Oswald that he had just visited Karungu and collected all the fossil material he could find, which had already been posted to London.

Oswald must have been bitterly disappointed. He had raised funds and traveled thousands of miles from London, only to find his main source of information had become crocodile food. Just when he had resurrected his hope of finding the fossil site by bumping into

Chesnaye, he learned that the fossils he sought were already on their way to London. Since he was already in Kenya, Oswald determined to search for the locality, which he found and studied. He determined that the geological deposits were fluviatile (river-laid) and he mapped several new fossiliferous areas. Since Oswald's detailed work on the paleoecology and stratigraphy of the Karungu site is still well regarded, the trip was not a total loss, though it surely was not the great success Oswald hoped for. Still, it must be said that the first two fossil-hunting expeditions in western Kenya (Pigott's and Oswald's) ended with whimpers.

Things were quiet then for a few years, but somehow the fossils of western Kenya just couldn't remain unnoticed. More Miocene fossils were found in 1927 by a former government medical officer, Dr. H. L. Gordon, on his land near Koru. Gordon was quarrying for agricultural lime at a locality that he called Maize Crib when he noticed some fossilized bones. Among them was a nodule containing an upper jaw; it was almost entirely encrusted with rocky matrix, but the tip of a long canine tooth was showing and he could guess what it was. Gordon sent this specimen and some others to E. J. Wayland, then director of the Uganda Geological Survey.

Edward James Wayland—Jimmy to his friends—was another of those inimitable types sent out to make good in the remote outposts of the Empire. Trained at the Royal School of Mines in geology, in 1918 he was named a geological expert and sent to become the first director of the Uganda Geological Survey. In January 1919, when he took up his appointment in Uganda, Wayland headed a staff of two: an assistant geologist, W. C. Simmons, and a clerk, E. N. Brohier. From that day until 1938, when Wayland left the survey, he never had more than five officers at his disposal at one time. His charge was disproportionate to the size of his staff; it was to survey the entire Uganda Protectorate (which then included much of what is now western Kenya) for minerals. While he was at it, Wayland was supposed to make a detailed geological map of Uganda, too. He was not in the least intimidated. Indeed, Wayland was a diehard field man whose first attempt at learning about the vast territory now in his purview was to walk around it for four months—a trip of some 1,200 miles—to see what was there. Needless to say, he had a tough constitution and a forceful personality.

Wayland's first assistant geologist, W. C. Simmons, did not last long. Simmons was sent out to work in Bunyoro, near Lake Albert. In Wayland's very first annual report, he wrote: "It is with much regret that one has to record the fact that the Assistant Geologist broke down in health at the end of November and was compelled to return, after medical treatment lasting over a month, to Entebbe accompanied by a medical man. After a short stay in hospital, Mr. Simmons was released for light duty. He proceeds to England in July, 1920, for leave and further medical advice." Though Simmons's physical deterioration was regrettable, it was not remarkable. Wayland reported that during fieldwork it was routine for 20 percent of his local helpers to be sick at any one time—and Europeans were thought to be more vulnerable to tropical diseases than Africans.

From the outset, Wayland understood that one of his chief duties was propaganda, to counter the "remarkable lack of interest in the Geology of Uganda that was found to characterize the European population as a whole, prior to the establishment of this department." Part of his job was to get the settlers to understand that geology might be of practical importance to them. In 1923, he founded the Uganda Literary and Scientific Society to try to promote intellectual interests in the colony, which could be a lonely place for a man of curiosity and learning. He was also instrumental in founding the *Uganda Journal,* which published original articles on anthropology, prehistory, geology, mammalogy, ornithology, tribal customs and languages, and just about anything else to do with Uganda. Some forty years later, when I arrived in Uganda as a starting professor at Makerere University, the *Uganda Journal* was still going strong. Its editor was kind enough to accept some of the first articles I published.

Like so many of the pioneers of East Africa, Wayland was a blunt man who tended to say exactly what he thought. Once the newly arrived governor of Uganda, Sir Bernard Bourdillon, asked Wayland if he thought it would be a good thing to bring out a "dowser" from England to find water; the governor was apparently no scientist. Wayland looked him straight in the eye and replied, in a quiet but deadly voice, that it would be better to hire a witch doctor on a temporary local agreement, and much cheaper. Sir Bernard was not amused.

Not surprisingly, Wayland thought the fossils from Koru were important. He encouraged Gordon to keep an eye out for more and to send any he found to him. A year or two later, Wayland packed up a selection of the better fossils from Gordon and sent them on to Arthur Tindell Hopwood at the BM for further identification and study.

Hopwood was a very tall, thin, rather quiet man, the sort who didn't enjoy the limelight but was perfectly happy burrowing away on some obscure problem in a dusty corner of the museum. What had come his way from Gordon via Wayland, he was delighted to see, was the jaw of an ancient ape. Very few ape fossils were known at the time, so this most chimpanzee-like creature was of supreme importance. After cleaning off the matrix adhering to the specimen and studying the fossil, Hopwood was convinced it was the ancestor of chimpanzees. When Hopwood eventually wrote up the specimen, he called the new species by the fanciful name *Proconsul africanus* to indicate its place in the phylogenetic or evolutionary tree of apes and humans. *Proconsul*, to almost anyone of that era, meant "toward the modern chimpanzee."

But were there more fossils of this fascinating creature? If so, they were in Kenya. Organizing a fossil expedition to Africa was no simple task in those days, as Oswald had already discovered. Hopwood had no African experience whatsoever and he was more a scholar than an outdoorsman. He sought the assistance of Louis Leakey, then a fellow of St. John's College at Cambridge University.

Even as a young man, Louis was legendary, for he was like no other Cambridge student before or since. The son of missionary parents, Louis had been born and raised in Kenya isolated from other Europeans save his family. As he grew up, his playmates and companions were Kikuyu, one of the most populous tribes of Kenya. Their language was his first language and he always said he dreamed in Kikuyu. While his parents did their missionary work, the children (Louis and his two older sisters) were educated by a series of spinster governesses. The net result was that the governesses taught Louis what *they* thought he should know and the Kikuyu elders taught him what *they* thought he should know. The sum total was unique.

Apart from his linguistic prowess in Kikuyu and some other African languages, Louis had a wealth of bushcraft that was of absolutely no use in England but that was essential to survival in the wilds of Africa. He also had an exceptional store of self-confidence that may well have come in part from being treated as an adult from the age of 14 on, as was traditional among the Kikuyu. Louis had few of the social skills of his Cambridge classmates, and none of their ever-helpful old school connections. His pre-university education had been irregular, unusual, and spotty. Nonetheless, it seems never to have occurred to Louis that he would do anything but succeed in the British academic system. As he did ever afterward, he simply plunged headlong into anything that interested him, assuming he could master it rapidly and ignoring whatever received wisdom there might be on the subject.

By the time Hopwood approached Louis to help him organize his trip to the eastern shore of Lake Victoria in search of Miocene deposits, Louis had started to organize an expedition to Olduvai Gorge, in Tanganyika (now Tanzania), for 1931. He knew there were fossils at Olduvai—a German, Hans Reck had found a human skeleton known as Oldoway Man there in 1913—and he hoped there were very ancient stone tools as well. In his total ignorance of the realities of European archaeology, Louis had been convinced since boyhood that the earliest habitations of human ancestors were in Africa. His view could be labeled "unpopular at the time" except that that is too generous an appraisal. Very, very few of his professors and fellow students expected him to find archaeological treasure in Africa, though Louis was proven right in the end. What he wanted to do was gather the hard evidence that would convince his skeptics. Donald MacInnes and Vivian Fuchs were to go along with Louis as geologists. Now Hopwood could be the official representative from the BM to the expedition, which sounded very grand. In exchange for Hopwood's tacit backing, Louis agreed to help Hopwood visit the Miocene fossil beds in western Kenya before going to Olduvai. At the time, Louis had little interest in the Miocene, which was much too old to yield tool-using human ancestors. Louis wanted to discover cultural beings.

In August 1931, Hopwood finally got to Kenya to see for himself where this ancestral ape had come from and whether or not there were more bits of it lying around. Once they all got to Nairobi, Louis's assistance consisted of lending Hopwood a lorry, helping him get together appropriate camping equipment, and finding him a driver-cum-mechanic. For the wage of £5 a month (then about $23,) this paragon maintained and drove the vehicle and translated among the various languages he spoke: Maasai (his native language), Kiswahili, plus two other local African languages, Hindustani, and English. Louis accompanied Hopwood and his driver to Koru, staying a few days to see that Hopwood's camp was properly set up before disappearing to conduct some excavations of his own much nearer Nairobi.

In the five weeks he spent in western Kenya, Hopwood recovered nine more specimens of apelike creatures from Koru. Some were *Proconsul,* like the original jaw, and some represented brand new species. There were also numerous other mammal fossils. There was a deinothere that was eventually named after Charles Hobley *(Deinotherium hobleyi),* some archaic carnivores of a type known as creodonts—crude, rather hyenalike creatures—various rodents, insectivores, pigs, and a few ruminants, or cud-chewing animals.

The primates were clearly the most spectacular finds. One of them was a vaguely gibbonlike species that Hopwood named *Limnopithecus legetet. Limnopithecus* meant "the lake ape"; the trivial or second name referred to both the hill on which Hopwood camped and the farm on which that hill stood, for both were called Legetet. The other new species he called *Xenopithecus koruensis,* the strange or exotic ape from Koru. His collection was an impressive success by any standards. When Hopwood had finished at Koru, he returned to Nairobi, well satisfied with his haul.

On September 22, 1931, the party set out for Olduvai over appalling, dusty, rutted roads and unmarked tracks that were even worse. The expedition consisted of Louis, Hopwood, a former Indian Army Captain J. H. Hewlitt, hired primarily for his shooting prowess, Professor Reck, and eighteen Africans. The paleontological and archaeological success that the expedition met with at Olduvai is legendary. The work at Olduvai continued for many years, but that tale is part of a story of stone tools and early human ancestors, not this story of fossil apes. At the end of 1931, Hopwood went happily home with crates of mammalian fossils from Olduvai and his precious, and even older, fossils from Koru.

By that time, Louis had already publicly proclaimed that Oldoway Man was the most ancient human skeleton in all of Africa. He suggested that the skeleton was contemporaneous with the very primitive stone tools he began to find almost immediately upon his arrival at Olduvai, but he found it difficult to convince his peers in Europe. The skeleton came from a geological level at Olduvai known as Bed II; Reck had been able to relocate the exact find-spot because he had stuck four wooden pegs into the sediments at the time that he found the skeleton. The key question was whether the specimen was a more recent burial dug into sediments much older than itself or whether the skeleton was contemporaneous with the bed. Louis had no doubt the skeleton was as ancient as could be, and he was supported by no less an authority than the anatomist Sir Arthur Keith, who had risen to prominence in heated debates over the correct reconstruction of the Piltdown skull, but there was still controversy.

When he got home to England, Hopwood tried to help resolve the problem by using a new technique. John Solomon, a geologist who had worked with Louis on a previous expedition, had begun using detailed analyses of the mineral content of different sediments as a geological tool for distinguishing different beds; this seemed a perfect case for applying this technique. Hopwood took home samples of the different beds at Olduvai for analysis by Solomon and then sent to Germany for a scraping of sediment from the original Oldoway skeleton. Unfortunately for Hopwood, who was trying to find support for Louis's idea, Solomon's analysis suggested strongly that the scrapings from the burial were from a different bed than the one in which the burial was found. In short, Solomon had convincing evidence that the skeleton of Oldoway Man was intrusive and certainly younger than Bed II, as is now widely agreed. Louis was deeply annoyed, for this verdict debunked the most spectacular find with which he had yet been associated. He never really forgave Hopwood for sending the samples out for an analysis that yielded such a disappointing result.

Slightly daunted, Louis decided he simply had to find another Oldoway Man, only somewhere where the stratigraphy would not be in doubt. He even wrote emphatically to Keith, who had by now joined the Oldoway Man skeptics, "I WANT IF POSSIBLE TO FIND ANOTHER OLDOWAY SKELETON." He set off to an area of western Kenya where an impeccable source, a missionary known as the Venerable Archdeacon Walter Owen, had been finding fossils. Owen had kindly invited Louis to visit and look over the deposits. Maybe in western Kenya he would find another very ancient skeleton of *Homo sapiens* that would convince his most ardent skeptics.

Accompanied again by Donald MacInnes and three other Cambridge students, Louis started prospecting in the badlands of Kanam and Kanjera, which were only a few miles from the shores of Lake Victoria. On their second day at Kanjera, March 14, 1932, MacInnes found some fragments of fossilized human skull washing out of the sediments. This was exactly the sort of find Louis had hoped for. More human fossils appeared in subsequent days. With this wonderful evidence in hand, Louis now backed off completely from the claim that Oldoway Man was the oldest human in Africa. Now that there were human fossils from Kanam and Kanjera, fossils that his own expedition had found, Louis could simply substitute them for Oldoway Man in his scenario of early humans in Africa.

As Louis's colleague John Solomon said, dropping Oldoway Man and replacing it with the man from Kanam and Kanjera made Louis look more like an "enthusiast" than a "scientist." A less kind assessment might have been that he looked like an opportunist determined to prove his ideas no matter how flimsy the evidence or how often he had to substitute one specimen for another.

This time, Louis thought, he had the goods and knew just where they came from. Piltdown Man from Sussex could no longer be considered the oldest man (that the Piltdown fossil was a forgery was not revealed until nearly 20 years later). Louis's fossil, Kanam Man, had ascended to the throne as the oldest man. In 1932, Louis announced his discoveries in the scientific journal *Nature.* A panel of prominent men—experts in anatomy, anthropology, and geology—convened on March 18, 1933, at Cambridge to review Louis's material and evaluate this claim. Louis gave a talk and displayed his stone tools and fossils, with his characteristic and engaging enthusiasm. The committee noted that they were "not able to point to any detail of the specimen that was incompatible with its inclusion in the type of *Homo sapiens,*" which left as the only question whether the fossil was genuinely ancient or whether it was a modern specimen buried intrusively in older beds. The committee explicitly asked for additional geological evidence, but on the whole thought Louis was to be "congratulated . . . on the exceptional significance of his discoveries."

In 1934, Percy Boswell, one of the eminent professors who had evaluated Louis's evidence at the Cambridge conference, came to follow up on the evaluation by examining the spot where the "new" oldest humans in Africa had been found. Boswell had always thought that the geological data Louis presented were inadequate, and he found Louis personally irritating. Despite his tremendous initial confidence, Louis was unable to relocate any of the sites with any certainty. Louis's golden success turned to lead: the fossils were good but the documentation was an abysmal failure. Cameras had malfunctioned, photographic film was fogged, and precise locations of the finds were not accurately enough recorded. Even Louis's final attempt to mark the locality failed. He had driven iron pegs into the sediments where the fossils were found, following Reck's procedure with—he must have thought—the improvement that he used long-lasting iron instead of wood. Louis had no good photographs, had no detailed measurements taken from a fixed point, and had no geological map. The iron pegs had vanished. Ironically, Louis's decision to use iron rather than wooden pegs was probably his last and most devastating mistake at Kanam. Metal was a far more precious and useful substance to people in western Kenya than wood. Undoubtedly, the pegs had been found and taken by a local, who probably forged spear points or a fine *panga* (machete) from the iron.

Boswell was appalled by Louis's carelessness, and Louis's heartfelt assertions that he could locate the site within "ten yards or so to my satisfaction" failed to convince Boswell. The entire occasion was embarrassing to them both. Louis's claims for the Kanam and Kanjera remains simply could not be supported by hard evidence about their in situ location because Louis could not find the site. When Boswell's assessment was published, Louis's reputation as a scientist was badly damaged.

To add to Louis's problems, rumors were now circulating—and they were true—that he was having an illicit affair. Some months before Boswell's visit, in January 1934, Louis had left his wife, Frida, his two-year-old daughter, Priscilla, and their newborn, Colin. He had left them not simply behind in England while he returned to Africa, but for good. He was passionately in love with a very clever and talented archaeological illustrator, Mary Nicol, who at 20 was some 12 years his junior. Adultery and divorce were then considered such scandalous behavior that it was many years before Louis lived down the consequences of falling in love when he was already married. The professional debacle at Kanam and Kanjera added feathers to the tar with which some colleagues were already prepared to cover him.

Louis was apparently unaware of the depth of Boswell's disbelief by the time the professor left his camp. Upon his return to England, Boswell attended several professional meetings, including one at the Royal Anthropological Institute, and told the story of Louis's utter inability to substantiate his claims. The juicy tale was repeated, with great satisfaction by those Louis had galled with his arrogant self-confidence and with sadness by his friends. Louis's failure was the talk of scientific London. Even his former supporters and friends, such as A. C. Haddon, were dismayed.

> So far as I can gather [Haddon wrote sternly], it is not merely a matter of a mistaken photograph, but a criticism of all your geological evidence at Kanam, Kanjera and Oldoway. The conference at Cambridge had to rely implicitly on your statements, and from what I hear there is much annoyance in view of recent developments. It seems to me that your future career depends largely upon the manner in which you face the criticisms. I am not in a position to know to what extent they can be rebutted by you with scientific evidence, but if you want to secure the confidence of scientific men you must act bravely and not shuffle. You may remember that more than once I have warned you not to be in too much of a hurry in your scientific work as I feared your zeal might overrun your discretion and I can only hope that it has not done so in this case.

The description of Louis as one whose zeal overran his discretion was probably very accurate.

For his part, Louis took the threat seriously, writing in his diary that "Boswell's findings may ruin my career." The paper that Boswell published about the business in *Nature* did little to reassure Louis.

Boswell painted Louis as sloppy in his fieldwork, hasty in his conclusions, and ignorant of basic geology. It was a brutal condemnation.

The only bright spot on the horizon was that, before receiving word of the reaction in England, Louis and his crew had visited Rusinga Island in Lake Victoria. They found some excellent fossiliferous sites, approximately contemporaneous with those at which Chesnaye, Pigott, Oswald, and Hopwood had collected nearby. In all, they recovered at least sixteen specimens of Miocene apes and over thirty species of other mammals from Rusinga and sites nearby on the mainland. This was not the earliest evidence of human beings in Africa, but it was something.

As Louis himself wrote,

> In the normal course of events I should not have spent much time at Rusinga, for, although I was interested in all kinds of fossils, I had my own work to do and could not afford to study strata that had no bearing upon the problems which I was concerned with. But during our first visit to Rusinga we discovered the remains of some Miocene anthropoid apes, and this immediately altered the situation. We believe that man and the great anthropoid apes are descended from some common simian ancestor, and such ancestor must have been present in some part of the world in Miocene times. . . . It therefore became imperative to study the Miocene deposits at Rusinga Island very carefully and to collect all the material that we could.

As I see it, the truth of the matter was that, having now found anthropoid apes himself, Louis was no longer disinterested. Suddenly Miocene apes—*his* Miocene apes—were urgently important. That was Louis's way.

The only sympathetic note Louis received over the Kanam-Kanjera debacle was from Hopwood, who perhaps felt a little guilty at his inadvertent role in dismantling the claim that Oldoway Man was the oldest man in Africa. "I was very sorry to hear about the results of Boswell's visit to the Kanam-Kanjera sites" Hopwood wrote, "and sympathise deeply on your disappointment. Doubtless your infernal luck will not desert you and you will confound your critics yet."

Louis certainly had infernal luck, both good and bad, but it didn't stop him from holding a grudge. In subsequent years, more and more Miocene ape fossils were found. Louis or Donald MacInnes revisited the Miocene deposits of western Kenya in 1932, 1934, 1935, 1938, 1940, and 1942, and often collected primate fossils. Hopwood would have been the logical and appropriate person to receive them for study, since he had already published on the first Miocene apes from Koru. Instead, Louis sent some of them to Arthur Keith, who was his mentor, his teacher of anatomy, and one of his few supporters over the Oldoway Man business. Keith erroneously thought they were the same as a known European fossil ape, *Dryopithecus,* not anything new. Sometimes Louis wrote up the finds himself or allowed MacInnes to report on them, but he never gave them to Hopwood. Hopwood had made his own luck by showing Louis up over the antiquity of Oldoway Man. He was not to be favored with the gift of Louis's precious fossils ever again.

2

Love and the Tree

Mary Nicol arrived in Tanganyika in 1935 to join Louis Leakey on a return trip to Olduvai shortly after the fiasco with Boswell at Kanam and Kanjera. Unfortunately, she came in the middle of the long rains, a time in East Africa when dirt roads and tracks turn into sucking, slithering, muddy quagmires and once-difficult travel becomes nearly impossible. Somehow it seemed romantic to them. They were in love, they were together in the wilderness, and they were going to work at Olduvai Gorge, where they were sure they would make great finds (and they did—eventually).

Once they got to Olduvai, the abundance of artifacts and fossils was stunning. Sam White, one of the students who had been with Louis at Kanam and Kanjera, was amazed. "What the hell were we doing messing around in the pittling little gullies of Kanam-Kanjera when this wonderland existed?" he puzzled. There was no real answer. Louis was always drawn in many different directions at once: Kanam, Kanjera, Rusinga, Olduvai, they all mattered to him.

Louis's European colleagues may have been sympathetic if not wholeheartedly approving of Louis's desire to bring Mary along, but his Kikuyu helpers were not. Louis's wife, Frida, had been on expeditions with these same men and they had liked her. White recalled that the Kikuyu asked him again and again if it was true that Bwana

Louis, the son of missionaries, would divorce his wife for another woman. To take a second wife was something a Kikuyu could understand; it was normal and even admirable for an African man who was succeeding in life to take a second or even a third, younger wife. But Louis was the scion of a family that had preached "one wife" relentlessly to the Kikuyus; the Leakeys had personally tried to persuade the Kikuyu that polygamy was a sin. For Louis to abandon his first wife and children in order to take up with another woman was unthinkable. But Mary was a small, neat, and formidably intelligent woman, careful in her work and happy in the field: hardly the prototypical "loose woman" who seduces someone else's husband. And, as her quiet competence at excavation and drawing became evident, even Heselon Mukiri, Louis's number-one assistant, began to accept her. Besides, the Kikuyu are practical people; Frida was gone and never reappeared, Mary was there and could do her job. The Kikuyu came around, but Louis's family was appalled and deeply grieved when rumors of Louis's behavior—rumors he could not deny—reached them.

When Louis and Mary returned to London to live in unwedded bliss, and to analyze the crates of artifacts and bones they had recovered from Olduvai, Louis had to face the signs that his career at St. John's was over. His fellowship was not renewed; he was offered no teaching jobs; and he was asked to vacate the rooms he had occupied. He and Mary moved to Great Munden, renting a little cottage where they lived with almost no money but great happiness. Louis was turned down for jobs and grants and fellowships repeatedly in the next few years, so complete was his ostracism from mainstream academia. That he and Mary married as soon as they were legally able, on Christmas Eve 1936, did nothing to change public opinion. He had behaved badly, both socially and scientifically, and this rejection by the closed circle of British academia was the consequence of his actions.

For more than 10 years after the embarrassment at Kanam and Kanjera, Louis and Mary worked doggedly at various East African sites, with no external funding but with considerable success. Rusinga and the Miocene beds of western Kenya were not the main focus of the Leakeys' work during that interval, though Louis and MacInnes paid a few brief visits to the sites. In 1946, Louis went back to England to show his *Proconsul* jaw around; he kept it on a cotton wool bed in a biscuit tin.

Louis's immediate aim in returning to England was to organize the 1947 Pan-African Congress of Prehistory, which was to be held in January in Nairobi. The organization was good, if carried out with little money or assistance, but it was the field trips that probably most impressed the delegates. Louis somehow took all sixty delegates from twenty-six different countries to see almost everything he and Mary had done or discovered: archaeological sites littered with hand axes at Kariandusi and Olorgesailie; fossil-rich Olduvai Gorge with stone tools as well; Mary's excavation of the Neolithic-Age site at Hyrax Hill and Louis's at Gamble's Cave. As a finale, the participants—who doubtless expected to see few fine art works in Africa—were taken to view the stunning, prehistoric rock paintings that Louis and Mary had discovered at Kisese. The average delegate's feeling must have been that Louis had been everywhere in East Africa. He had excavated sites ranging from millions of years old to a few thousand years old. If there was anything of interest between the beginning of humankind and yesterday that Louis had missed—and there was of course, but no one knew it then—its absence was overwhelmed by the vast riches he had already uncovered.

Attending the conference was Wilfrid E. Le Gros Clark, known simply as Le Gros to most of his friends and colleagues. Le Gros was a tall, quietly impressive man, with a rich, plummy voice that signaled his very well-to-do origins. He was not simply well born; he was the possessor of a careful intellect and an exhaustive knowledge of primate anatomy. At the time, Le Gros was the professor of anatomy at Oxford and probably the most respected anatomist in all of Britain, if not the Western world. Unlike many of the others at the conference, who were dazzled by the human fossils and cultural arti-

Louis Leakey (right) and Sir Wilfrid Le Gros Clark at Oxford when Louis received an honorary degree. (Copyright The Leakey Family Archives.)

facts of Africa uncovered by Louis and Mary, Le Gros was fascinated by the Miocene apes. He had seen Louis's *Proconsul* jaw in England and probably Hopwood's material as well; now he had the chance to examine all of the other primate fossils from Rusinga, Songhor, and elsewhere in western Kenya that Louis had been accumulating for years. At the congress, both Louis and MacInnes gave papers about the Miocene sites.

Le Gros's viewpoint was different from that of most of his colleagues in at least one important respect. Just before the Pan-African Congress, he had visited South Africa and examined the australopithecine finds made by Robert Broom and Raymond Dart in the 1920s and 1930s. From 1925, when Dart reported the first fossil skull, he had been convinced he had found a human ancestor. But Dart was an emotional man, an Australian anatomist who wrote about human origins in purple prose with an almost evangelical flavor that made his colleagues doubt his objectivity. Broom, a stubborn old Scots physician and paleontologist, was one of the few who agreed with him. Hardly anyone from the European academic establishment had traveled to look at the South African specimens firsthand; australopithecines were largely dismissed as being too apelike to be human ancestors. But when Le Gros examined the material for himself, he realized that these were indeed very primitive hominids (members of the human zoological family) and the oldest known ancestors of modern humans. Dart and Broom were right. Le Gros announced his radical conclusion at the Pan-African Congress. He was so well respected that his endorsement of the australopithecines as hominids sparked a massive rethinking of the consensus on how and where humans had evolved.

As early as 1871, Darwin had written:

> Where was the birthplace of man at that stage of descent when our progenitors diverged from the Catarrhine [Old World monkey] stock? The fact that they belonged to this stock clearly shows that they inhabited the Old World, but not Australia nor any oceanic island. . . . In each great region of the world the living mammals are closely related to the extinct species of the same region. It is, therefore, probable that Africa was formerly inhabited by extinct apes closely allied to the gorilla and chimpanzee; and as these two species are now man's nearest allies, it is somewhat more probable that our early progenitors lived on the African continent than elsewhere.

Now Dart and Broom had found very early, very primitive human ancestors in South Africa. This knowledge transformed Africa from an obscure, out-of-the-way place that had had little to do with human origins to a very likely cradle of humanity. It was certainly a novel idea that was well worth exploring. The Miocene sites where

Louis had already found apelike fossils—fossils that might in turn be ancestral to australopithecines—were an obvious target for further research.

Louis and Mary took a few of the conference delegates on yet another excursion, this time to Rusinga. In addition to some younger, fitter scientists, the party included Miss Dorothea Bate, a renowned lady archaeologist who specialized in analyzing mammalian faunas, the fossilized animal bones found at early living sites. In her youth, Miss Bate had had much experience on digs in the Middle East and had worked on Neanderthal sites at Mt. Carmel in the Levant. In 1947 Miss Bate was a venerable lady of ironclad propriety that had not been shaken or diminished by her travels. She had some difficulty making the descent into Olduvai Gorge on this journey, but that was hardly surprising considering she was about 70 at the time. She was still game when the opportunity arose to go to Rusinga and see the Miocene sites.

According to an account by Mary Leakey, the motor launch lent to the Leakeys for the expedition "was put gently but firmly onto a submerged rock by an inexperienced helmsman and stuck fast." Running aground looked like a minor embarrassment, merely a mildly amusing inconvenience, until the question of how the party was actually to get ashore was considered.

"Many of the islanders," wrote Mary, "who were Luo, swam or waded out to help us. Some were old friends of ours, but all were undeniably completely naked, and Miss Bate had to cover her eyes. Worse was yet to come, for the boat in its laden state could not be lifted off the rocks and we were faced with wading ashore or being carried on the shoulders of the cheerful and still naked Luo men. Miss Bate was not prepared to wade."

It is a wonderful image, Miss Bate with her Victorian sensibilities being carried to the shore in the strong arms of a soaking wet, naked, black Samaritan. I imagine her shuddering slightly at the intimacy of the contact, with her eyes firmly closed. I never knew Dorothea Bate, but I can't help wondering if she wasn't concealing a certain glee at the adventure of it all. Perhaps she felt she was expected to give a horrified response and obliged. Many of the female scientists of the early part of the twentieth century were much tougher and more pragmatic than they looked. Certainly Miss Bate was entirely used to working in the field side by side with scores of Arab men, if not with naked Africans.

In any case, the field trips and the congress itself proved most successful at restoring Louis's reputation and provoking a new focus on African fossils. Le Gros left the conference committed to helping Louis find serious funding for a Miocene expedition. Le Gros was not a field man himself, but he was impeccably well connected and very kind. Within weeks Le Gros had approached the Royal Society, a bastion of establishment academia, which ultimately gave Louis £1,500 (equivalent at the time to about $6,000) for Miocene research. Louis, for his part, approached His Highness the Aga Khan, the leader of the Ismaili Muslim sect who lived in Kenya and was, fortunately, interested in prehistory. The Aga Khan donated £250 (about $1,000 in those days) for the Miocene work, some camping companies gave a few tents, and the British Overseas Air Company (the forerunner of British Airways) gave Louis one free airline ticket. After 10 years, Louis had been rehabilitated as a professional with Le Gros's help.

The British-Kenyan Miocene Expedition to Rusinga was launched in a rush. To his horror, Louis had seen some advance press from the University of California Africa Expedition, led by Wendell Phillips from Berkeley. Phillips had raised huge sums of money, especially by Louis's standards: over $150,000, plus vehicles, trucks, camping gear, 10,000 gallons of gasoline, and round-trip tickets for twenty scientists. His group planned a massive Cape-to-Cairo expedition with the intention of resolving all of the problems of human prehistory, or so they hoped. Louis tried to be civil and offer assistance to Phillips, but he was understandably a little nervous that his years of work prospecting for sites in East Africa were about to be usurped by a rich American.

Louis was greatly relieved when he, Mary, and their two boys, Jonathan and Richard, set out for Rusinga Island on July 8, 1947. There was as yet no sign of Phillips's expedition. The only other Europeans going into the field with them were the paleontologist Donald MacInnes, and two geologists, Robert Shackleton (first cousin to the famous Ernest Shackleton of Antarctic exploration) and Ian Higginbottom. Heselon Mukiri, Louis's most skilled excavator, headed a small team of Kenyan workers, all of the Kikuyu tribe. By Louis's usual standards, it was a luxury safari. They had enough money for a car, a rented truck, and a rented motorboat, called Maji Moto, or Hot Water. The Maji Moto made getting to Rusinga Island from the mainland at Kisumu only a five-hour trip, if the weather was good. Now there is a causeway connecting Rusinga to the mainland and the trip is much faster.

Though the Kavirondo region is inhabited by people who rely heavily on boats and fishing, they do not take Lake Victoria lightly. The lake is an enormous body of water that seems more like a huge inland sea than a lake. Victoria is notorious for the ferocity of its thunderstorms and the unpredictability of its surface. Glowering purple-black skies that are split by crackling lightning bolts are quite common, as are water spouts: the aquatic equivalent of tornadoes. No one, not even the local Luo fishermen who grew up there, is safe venturing onto the lake in inclement weather.

The expedition camped under a lush fig tree near Kiwegi Hill, a small peak near the eastern shore of Rusinga Island that is capped with lava from the ancient volcano that formed the island. Louis later recalled that no fewer than three pairs of fish eagles used that same tree as their habitual roost. Fish eagles are lovely birds similar in appearance to American bald eagles. Their high, keening cry is oddly soothing, one of the nicest memories I have of camping on Rusinga. Unlike Olduvai or many of the other East African sites where Louis and Mary worked, Rusinga was green, thickly vegetated, and wet. In the rainy season, "wet" is an understatement. The land becomes a morass and the single dirt road that winds its way around most of the island's perimeter turns into a muddy stream. In those days, the island was rather thinly inhabited by the Luo people, who fished and swam, sang cheerily, and talked loudly while they grew millet and maize and produced plump black babies. In 1947, Louis and Mary went to dig on a densely wooded island, with a few cultivated fields interspersed between ancient stands of trees.

By 1984, when I got there with my team, Rusinga was the island of the schoolchildren. In the mornings, the surface of the single road could not be seen; it was completely obliterated by hordes of schoolchildren in uniforms once bright pink, green, and blue but faded by frequent washing. In my day, the island that had been covered in trees was a patchwork of small farms, or *shambas;* only a very few trees clung to the steepest slopes and the most difficult parts of the island to reach. Human population growth had taken its toll on the habitat. There were still a few large fig trees inhabited by fish eagles along the shore and, like the Leakeys, we camped under them happily.

The British-Kenyan Miocene Expedition had an amazing first field season in 1947. By the end of it, they had collected some 1,300 fossils, including the apelike primates that were already known, most of the skeleton of a rhino related to the Sumatran rhino of Asia, and a giant hyrax. Hyraxes are today small furry creatures that look somewhat like short-eared rabbits or small groundhogs. They live on cliff faces or in forests, depending on whether they are rock hyraxes or tree hyraxes. Although they are charming little animals, at night tree hyraxes make ungodly noises, like a woman being strangled. The ancient hyrax was as large as a calf. My imagination fails me when I try to conjure up the calls giant hyraxes must have vented into the dark Miocene night.

There were also pigs and lizards and plenty of those strange, primitive-looking Miocene species that are no longer alive today: clumsy-seeming, heavily-built carnivores; a large, hippopotamus-like creature known as an anthracothere; and primitive relatives of antelopes and giraffes.

The best fossil of all was found the third day. Louis and Mary had gone by boat around to Kiahera Hill, where Louis had found the *Proconsul* jaw in 1942 that had helped attract Le Gros's attention at the Pan-African Congress. Within minutes of arriving at the site, Louis picked up the canine tooth of the very same individual, making the specimen more complete. The island's fossil deposits seemed to hold a sort of Miocene bestiary, an extraordinary collection not only of extinct animals, but also of long-gone plants, seeds, snails, insects, and fish. It was a scientific treasure trove.

In the subsequent, 1948, field season, Louis received generous funding from a new benefactor, Charles Boise. It was then that Rusinga yielded one of the most spectacular fossils that the Leakeys found in a lifetime of noteworthy discoveries: a skull.

Mary made the find. She once wrote, "I would never claim that the Miocene, then or now, was one of my major interests. . . . As for the fossils themselves, many seemed to me merely boring and repetitive while others were among the most exciting things I have ever seen come out of the ground." In her heart, Mary probably wanted to find more archaeological sites full of beautiful stone tools that she could study. She loved stone tools, but she wasn't going to find these on Rusinga. What she found was something much older and much more important.

On the second of October, 1948, the expedition was camped on the northwest side of the island, near Kaswanga Point. Louis was busy excavating a crocodile at a site known as R106. (The R stands for Rusinga, 106 for the one-hundred-and-sixth site to be found there.) To use her own words, Mary had "never cared in the least for crocodiles, living or fossil," so she wandered off, looking at exposed cliffs and eroding land surfaces, to see if she could find something a little more appealing. She came upon a scatter of bones eroding out of a small cliff, under a tree. Her eye followed the trail of fragments up the slope and recognized something worth looking at more closely. It was a tooth. *No,* she realized as she brushed gently at the fossil, *it was not a tooth; it was a tooth in place in a jaw, and an-*

Heselon Mukiri, Louis Leakey's right-hand man in his early excavations on Rusinga and through all the years that Louis and Mary worked at Olduvai Gorge. The species *Proconsul heseloni* is named for him. (Copyright The Leakey Family Archives.)

Heselon Mukiri works under a shade umbrella excavating the 1948 skull of *Proconsul heseloni* that was found by Mary Leakey. The distinctive tree above him has now been cut down, but continues doggedly to send out new shoots. (Copyright The Leakey Family Archives.)

other tooth, and . . . She shouted to attract Louis's attention and he left his crocodile to see why she was calling. They soon realized they had not only a tooth, not only a jaw, but the better part of a skull of one of the Miocene apes. Mary later wrote: "This was a wildly exciting find which would delight human paleontologists all over the world, for the size and the shape of a hominid [*sic*] skull of this age, so vital to evolutionary studies, could hitherto only be guessed at. Ours were the first eyes ever to see a *Proconsul* face." Mary was not one to exaggerate her emotions, even well after the fact, which was when she wrote these words. She also confessed that the joy of the find led to another life-changing decision. She and Louis celebrated back at camp that night in the time-honored fashion of lovers and decided that having another baby would be a good idea. Nine months later, their third child, Phillip, was born.

Over the days that followed, the foreman, Heselon Mukiri, painstakingly brushed and chipped the sediment aside grain by grain while Mary sat nearby and received each fragment as Heselon freed it. His meticulous excavation and careful sieving of the backdirt yielded roughly thirty fragments of a skull and jaw of *Proconsul*. Mary had an excellent eye for a puzzle and spent many hours assembling these pieces into a largely complete skull, the first ever seen of a Miocene ape. Louis and Mary knew this was a stunning discovery, one that would make the headlines in the newspapers and in the scientific journals. A skull could yield crucial information about brain size, face shape, and more about the animal's teeth and diet. A skull was a marvel of information. The specimen itself is an exquisite little object, an ancient face with orbits that seemed to look out of the past.

"WE GOT THE BEST PRIMATE FIND OF OUR LIFETIME," Louis cabled Le Gros as soon as he reached Nairobi. On October 31, only a few weeks after its discovery, the little *Proconsul* skull was on its way to London, packed neatly into a biscuit tin that Mary held nervously on her lap the entire way. She was greeted by reporters as she left the BOAC plane at Heathrow and filmed for a press conference. Fortunately for Mary, who was then a rather shy woman unused to such attention, she was able to show the skull and not say much. Louis had already primed the press with a statement that the fossil showed "resemblances to the human condition," by which he meant that it lacked the large brow ridges and bony crests commonly found on the skulls of male chimpanzees and gorillas. Louis liked all of his important fossil finds to be direct human ancestors and the *Proconsul* skull was no exception. On November 8, he wrote to Le

Mary Leakey shows the new *Proconsul* skull to the press at London Airport in 1948. (Copyright The Leakey Family Archives.)

Gros explaining his view of the significance of the skull: "To me, this particular species of *Proconsul* appears to be a creature which could easily be in the direct ancestral line leading to man, and also possibly to some of the apes, as it is so generalised. . . . For a good many years now I have expected that the early members of the ape-human stock would have much more simple and less specialised features."

Though he never lacked for opinions, Louis's most crucial reason for sending the skull to London with Mary was so that Le Gros could help describe and analyze the skull and the other primate fossils from East Africa and write them up in a scientific monograph. Their work was published in 1951 as Clark and Leakey, *The Miocene Hominoidea of East Africa*, by the British Museum (Natural History). A hominoid is a member of the zoological superfamily that encompasses the living apes and humans and other extinct descendants from their last common ancestor. The skull was numbered R.1948, 50 (R106)—a

Rusinga find made in 1948, the fiftieth specimen, from site R106—and joined the portion of the right upper jaw and teeth that Hopwood had designated the holotype of *Proconsul africanus* in 1933. They were both considered part of the British Museum's collections, an issue that would become inflammatory years later.

While Louis was a good field man and full of enthusiasm for the big picture, he was not one to write a meticulous, exacting, ordered, point-by-point description of the skeletal anatomy of a specimen, followed by a rigorous and thoughtful comparison to other specimens of similar species. Compared to Le Gros's expertise, Louis's anatomical training was decidedly thin. Louis flatly needed Le Gros, or someone like him, to write the monograph. Reading the work now, with its measured tones and cautious conclusions, I believe Le Gros wrote nearly all of it, while Louis contributed information about the discoveries and their field context. Even the field information was phrased in exceedingly cautious terms, with specimens that had been collected on the surface, about which "there necessarily remains an element of doubt concerning its exact origin," painstakingly distinguished from those collected in situ. The sites themselves were mapped and numbered, not loosely named, and there is a complete listing of Kenyan sites from which Miocene fossils have been recovered. They wrote, "While there seems no reason at present to alter the provisional estimate of early Miocene [age], it should be pointed out that the material includes some [nonprimate] genera which, in other parts of the world, extend their range to horizons that are considerably younger than the Lower Miocene." As for the exact chronological relationship between the sites at Rusinga, "We have not attempted to make any stratigraphic correlation of the primate species, for the reason that the excavations in the area are still in progress and it is anticipated that much more material will be collected in the next five years."

Louis's flamboyant turn of phrase and grand claims are rarely evident in the monograph. Instead, the dominant voice is Le Gros's, calm and scholarly. There would be no repeat of the debacle at

Kanam and Kanjera, where much was claimed and too little was substantiated.

Notwithstanding Le Gros's conservatism, the monograph presented some astonishing facts and observations. First of all, the authors identified not one but six species of higher primate in the collection, four of which had been announced as new species in a short paper by Le Gros and Louis that preceded the monograph by a few months. The higher primates included three species of *Proconsul* (*P. africanus, P. nyanzae,* and *P. major)* that represented small-, medium-, and large-sized creatures; and two species of *Limnopithecus, L. legetet* (the gibbonlike species Hopwood had named from the Koru specimens), and a new one, *L. macinnesi,* from Rusinga. Even for species that had been known previously, there was a wealth of new material to be described, analyzed, and shown in this monograph.

There was also a primate upper jaw that was named *Sivapithecus africanus* by Le Gros and Louis, which was a new species in a genus formerly known only from India. The trouble was that no one was very sure where the specimen actually came from. MacInnes, who collected the specimen, never gave a precise locality. Le Gros and Louis assumed the jaw came from Rusinga, as did the other specimens collected on that trip, but the fossil was a peculiar color for a Rusinga fossil. An analysis of the matrix still adhering to the specimen in 1979 showed conclusively that the jaw was not from Rusinga. Its general preservation and the mineral analysis show that the specimen might come from Maboko, another island in Lake Victoria at which MacInnes excavated in 1934. But even if we discount the so-called *Sivapithecus* specimen, there were an unusual number of fossil primate species from Rusinga and a few other sites in western Kenya.

One of the first questions to arise about this collection of fossil primates is, *What are they are doing there?* The study of fossilization and preservation of bony remains, known as taphonomy, was not even in an embryonic stage at the time, so the best that could be done by way of explanation was to offer an informal hypothesis. Because the bones were invariably fragmented, and were sometimes found in sediments that Louis thought were lacustrine, or lake-deposited, either Louis or Le Gros wrote in the monograph: "It seems possible, although by no means certain, that the hominoids of this period were very liable to attacks of carnivora. The presence of the remains of so many individuals in shallow-water lacustrine deposits suggests, perhaps, that they were particularly vulnerable when they came down to the lake edge to drink, and that they were attacked and killed in large numbers under these conditions, so that the broken remains were washed into the lake beds from the adjoining shore at very frequent intervals."

There are echoes of the story of Dayrell Botry Pigott here and the anecdote was a favorite of Louis's. It is tempting to speculate that Louis drafted this section and Le Gros inserted all the cautions. We now know that nearly all of the primate fossils from Rusinga come from ancient soils, not lake beds.

Another key issue that the monograph brought out was the question of how different species of a similar type of animal were to be distinguished from one another. Most of the fossil specimens were loose teeth, or teeth in partial jaws or palates; this is a fairly common circumstance, since teeth and the bone surrounding them are not edible and are unusually dense: thus jaws survive well. But suppose, as was the case here, there is a large array of teeth that seem similar in anatomical features and in morphology, or shape, but variable in size. How much does a single species differ in size from individual to individual? This is a central problem in paleontology and always has been. The abundance of primate specimens found together on Rusinga, especially, underscored the difficulty of sorting out such complex assemblages of fossils.

Le Gros took the sensible first steps. He measured the teeth in various dimensions and constructed ratios of one tooth or one segment of the tooth row to another—for example, of the nipping incisors and canines at the front of the jaw relative to the chewing battery of premolar and molar teeth. He also relied on what might be called informed comparative intuition. He characterized *Proconsul africanus* as "intermediate in size between a large gibbon and a chimpanzee. . . . Mandible strongly built but small"—with no mention of the size of its canine teeth—while *P. nyanzae* was a "species of *Proconsul* approximating in size to a chimpanzee . . . Canines powerfully developed . . . Mandible large and strongly built." The quandary was complicated further by the presence at Songhor, another site in western Kenya, of *P. major*, another closely related species that seemed more massively built still.

What, exactly, is a strongly built but small mandible, or lower jaw? a pedant might ask. When does a mandible qualify as "large"? When does it become massive? Good questions: difficult answers.

It is important to understand that Le Gros's descriptions and analysis were neither deficient nor foolishly incomplete by the standards of the day. For one thing, the complex mathematical and statistical techniques for quantifying and comparing the shapes of three-dimensional objects—a field known as morphometrics—were still being developed in the 1950s. Even the morphometric techniques that did exist at the time were rarely used, in part because electronic calculators and computers were not readily available. Few practitioners of anatomy or paleontology were concerned about the variability within a species yet, so they tended to ignore the statistical techniques and problems. The painstaking slog of quantifying the variability in extant animals, so that you could get a precise measure of how variable the individuals in a single species might be, had not yet been undertaken for most species. A chimpanzee was . . . well, chimpanzee-sized, more or less. No one could tell you, with assurance and authority, that small and large individuals of *this* type of species typically varied by *this* much in dental size, bone size, or bone or tooth shape—never mind the additional confusion caused by difference in size and shape between the bodies and body parts of males and females.

Primitive as their techniques look by twenty-first-century standards, Le Gros and Louis got the basic message of the Miocene primate fossils right: There were a lot of hominoid species, apparently living together (or near enough to each other to be fossilized together) at one point in time. This evolutionary pattern of a proliferation of roughly similar types of closely related animals in one region is a prime example of adaptive radiation. The adaptive radiation of hominoids in the Miocene of Africa was genuinely different from the modern distribution of apes.

African hominoids are represented today by the gorillas and the chimpanzees. Most modern genetic and morphological studies suggest that there are more African apes than we once thought. There are at least two gorilla species—*Gorilla beringei* from the mountains and lowlands of Rwanda, Uganda, and the Congo, and the West African *Gorilla gorilla*—and at least two chimpanzee species—the bonobos, or so-called pygmy chimps, and the common chimpanzee. Some scientists maintain that the genetic differences between the common chimpanzees of eastern and western Africa are sufficient to consider them two separate species; doing that would make a total of three species of chimpanzees.

But these numbers of modern apes do not compare with the the numbers in the Miocene record, especially when you realize that the two gorilla species are geographically separated from each other and so are the two or three chimpanzee species. In zoological terms, these species are not sympatric—which means literally "living in the same country"—with the other species of their type. A similar pattern holds true in Asia: one species of orangutan does not live sympatrically with another. Sometimes one gibbon or siamang species overlaps with another, and then there are important ecological and behavioral differences between them. Yet all of these Miocene apes not only lived at the same time, but were also sympatric or close enough to it that all were fossilized within a circumscribed region.

Why is geographic separation from like species important? Isolating mechanisms, including geographic barriers, prevent direction competition between closely related species for resources such as food, water, breeding sites, and nesting sites. Isolating mechanisms also play a major role in the evolution of two or more species from a single parent species. Suppose a single ancestral population of hominoids were subdivided into two or three populations by some sort of barrier that prevented interbreeding. Geographic barriers are a clear example, but they are only one of the most obvious of many sorts of barriers to interbreeding, which include differences in breeding season, differences in mating calls, behavior, or coloration, and simply failure to recognize another individual as a conspecific (a member of the same species) of the opposite sex. The existence of such barriers to interbreeding means that mutations arising in one population would not be passed to the other population. Over time, the isolated populations could become progressively more different from each other. If those differences became so great that successful interbreeding was impossible even if the populations were reunited, there would be two species where once there was one.

What this brief review of the current number and distribution of apes shows is that it was unusual to find two *Proconsul* species and a somewhat gibbonlike *Limnopithecus* species in one site. The coexistence of several ape species seemed more extreme in the monograph than it actually was, because Le Gros and Louis tended to treat all of their specimens, from Rusinga and the mainland alike, as if they came from a single site.

Where we might expect (by analogy with modern conditions) to find geographic separation among ape species, in the Miocene we find instead coexistence and sympatry. If there is no geographic isolation, then there is only one way all of these Miocene primates could live sympatrically on Rusinga. The only explanation that makes sense in terms of evolution is that those ape species were separated by ecological differences that prevented them from competing all the time for the same resources. Such differences might have taken the form of different adaptations for feeding (diet) or moving about the world (locomotion) and those differences ought to be visible in the fossil record. What was needed in order to see those differences were techniques for discovering how extinct animals functioned, for deducing life habits from dead fossils.

Another crucial point that came out in the monograph by Le Gros and Louis was that these Miocene primates were not simply hominoids in the sense that modern apes are hominoids. A hominid (or, in a terminology becoming increasingly common, a hominin) is a subdivision of the hominoid superfamily that includes only humans and their immediate ancestors after our evolutionary path diverged from that of the great apes, the other main subdivision of the superfamily sometimes referred to as pongids.

Those categorical definitions seem very simple and very useful in the abstract. The big question, as Louis and Le Gros saw it, was: *Were these fossil primates monkeys or apes?* But the question itself is misguided.

Today paleontologists recognize a psychological phenomenon called "the pull of the present," which is almost always at work when you are dealing with fossils. What this term refers to is the tendency to think of the current situation as normal, to use the present as a lens through which we peer darkly at the past. Because we know only modern apes, we tend to take their traits as diagnostic and typical of all apes that ever lived. This habit of thought leads us to believe that an ape *must* be like either a gorilla, a chimpanzee, an orangutan, a gibbon, or a siamang because those are all the living creatures that come under the heading "ape." Even when we are dealing with apes that lived millions of years ago in the Miocene, we are in danger of unconsciously equating them with modern apes as if they had, simply through some piece of bad luck, become extinct.

The fallacy in this confusion of "ape" and "modern ape" lies in the fact that evolution is ongoing and continuous. Modern apes have had millions of years to evolve and change since the Miocene—so of course modern apes don't still look like Miocene ones, nor did ancient apes have all the features of modern apes.

In taxonomic terms, this error comes from confusing a crown

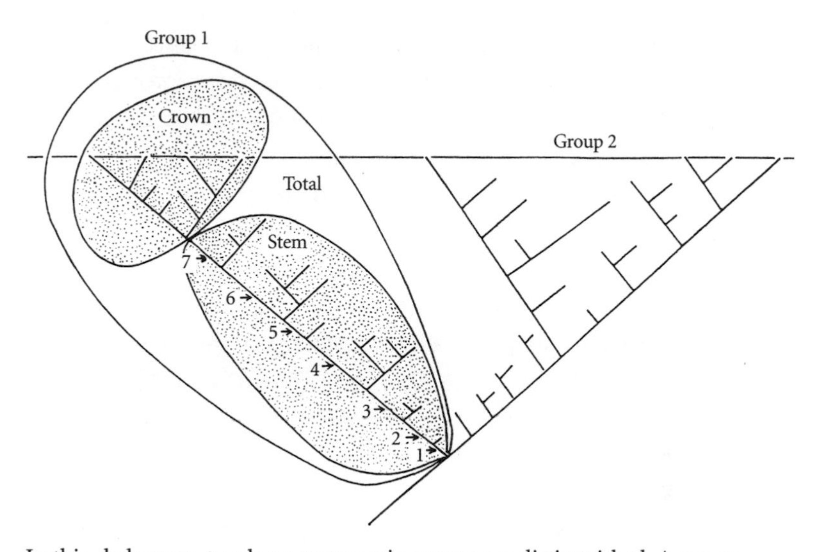

In this cladogram, two large taxonomic groups are distinguished. A crown group (within a larger lineage, labeled Group 1) is comprised of living species—those intersected by the horizontal line indicating the present—and all their extinct relatives back to the most recent common ancestor. The stem group is all the species (numbered 1–7) in that same lineage up to the most recent common ancestor of the crown group. (From R. P. S. Jefferies, *The Ancestry of Vertebrates* [London: British Museum (Natural History), 1986], p. 12; © The Natural History Museum, London.)

group with a stem group within an evolving lineage. A crown group is a group of species united by the possession of a common and major adaptation that nothing before them in their lineage possessed. Crown groups usually consist of living and very recently extinct species. In contrast, the species that occupied the beginning of the lineage is known as the stem group. The first ape to evolve was a stem ape; the last to evolve is a crown ape. It is clearly unrealistic to think that the very earliest members of this ape-and-human lineage (the stem hominoids) will look like the evolved, living ones (the crown apes).

Instead of being modern apes—crown apes—the Miocene species were primitive stem apes that were more monkeylike and less apelike than apes today, if "apelike" is defined as having all of the characteristics of living apes. What Miocene apes really were was just that: apes of the Miocene, apes that had evolved to be different from monkeys but were not yet what apes are today.

The distinction is perhaps best explained by a thought experiment. Suppose one of these Miocene hominoids were discovered alive on a secluded tropical island or, more sadly yet, were captured alive and sold at a remote meat market in some poor country. Suppose a biologist saw this creature and, realizing it was an animal new to science, bought it, saved its life, and studied it in detail. What would we call this new animal? Almost certainly, we would call this new animal an ape at first glance and maybe at second glance, too. We have no word other than "ape" that is more appropriate for such a creature in our common vocabulary even though modern apes are as distant, evolutionarily, as humans are from those Miocene apes.

If there was any doubt, the suite of Miocene hominoids from Kenya proved that these ancient hominoids had not yet developed all of the anatomical characteristics that we can observe in modern hominoids. This is no surprise, if you think as I do that evolution actually happens, but it is amazing how often people forget to allow for evolution to occur *over time.* Thus while we may be descended from apes, we are not descended from modern apes; no chimpanzee's daughter is ever going to walk out of the jungle as a human and no *Proconsul* daughter ever did. If *Proconsul* was our direct ancestor, then at least one million generations separates it from us. The primitiveness of *Proconsul* and its kin reinforces this perspective on our own ancestry.

As for anyone walking out of the jungle, the Miocene fossils from Kenya gave some tantalizing hints about that, too. Le Gros and Louis's 1951 monograph also discussed some fossilized primate limb bones that had come to light during the excavations. Unfortunately Louis was not well versed in primate anatomy and Le Gros had little familiarity with other species represented at the site. There were two foot bones from the site of Songhor, attributed incorrectly on the basis of size to *Proconsul nyanzae* and two partial clavicles, sent to Le Gros by Louis as primate, but these were actually crocodile femurs. On the basis of these and other specimens, Le Gros and Louis thought that *Proconsul* moved mostly on all fours—quadrupedally—and probably on the ground. They commented:

> If *Proconsul* occasionally raised itself on its hinder extremities like the modern Pongidae, it was able to throw its weight back more towards the heel, and thus to balance itself in the erect posture more effectively. The possible implications of this in relation to the subsequent development of a bipedal gait [as in humans] are sufficiently obvious. It is not suggested here that any of the early Miocene hominoids from East Africa necessarily bear a relationship, direct or indirect, to the ancestral stock from which the Hominidae originated. But at least it seems probable from the conformation of [these foot bones] . . . that the tarsal pattern of the human foot might have been more readily derived from that of *Proconsul* than from that characteristic of any of the modern large anthropoid apes.

Why isn't *Proconsul* just a monkey? The weightiest piece of evidence lies in its teeth. Anatomists divide teeth into different types, according to their placement in the jaw, their shape, and their embryonic development. If you look in the mirror for a moment, you can count your teeth from the middle front of your mouth backward, using either your upper or lower jaw. If you haven't had some teeth removed in your lifetime, and if you don't have a genetic variant that affects the number of your teeth, you will find that you have two flat, spatulate incisors at the front of your jaw, then one pointy canine, or eyetooth, then two premolars, or bicuspids, then three large, multicusped molar teeth at the back. This usual array of teeth is expressed as a dental formula that looks like a fraction, a sort of shorthand for indicating the dentition typical of a species; any tooth typically found in the upper jaw is written above the line, and any in the lower jaw below the line. Humans, apes, Old World monkeys, and the ancient Miocene apes all have the dental formula $\frac{2.1.2.3}{2.1.2.3}$. Monkeys from the New World have an extra premolar in both the upper and lower jaw, which gives most of them a dental formula of $\frac{2.1.3.3}{2.1.3.3}$. A higher primate with a $\frac{2.1.2.3}{2.1.2.3}$ dental formula could thus be an Old World monkey, an ape (modern or Miocene), or a human; the dental formula alone will not tell you who is who.

The shape of the teeth and the placement of the cusps yielded more clues. All living Old World monkeys have molar teeth that are bilophodont, meaning that the teeth have four cusps connected in pairs, or lophs—two at the front of the tooth and two at the rear. No higher primates other than Old World monkeys have bilophodont teeth, though, interestingly, kangaroos, tapirs, and some other animals quite distantly related to primates do. Fortunately, tapirs and kangaroos and other bilophodont animals lack other primate characteristics and so do not confuse the issue.

The practical rule is that identification proceeds from the grosser and more obvious traits to the more subtle ones. If you have a primate, and the number of teeth tells you it is an Old World primate, then the disposition of the cusps on those teeth may be what tells you if you have a monkey or an ape. These characteristics do not answer the bigger and more important question—*what really makes an ape an ape, or a monkey a monkey?*—they give you only a simple means of classifying specimens into preexisting categories.

In the tree of life, *Proconsul* clearly lay between monkeys and the inclusive group of apes and humans. The size gradations in *Proconsul* made it tempting to draw a branch of possible descent between *P. major* and gorillas, another between *P. nyanzae* and common chimpanzees, and a third that Louis especially favored: a branch that began with *P. africanus* and culminated in humans.

Was he right? And how would we know?

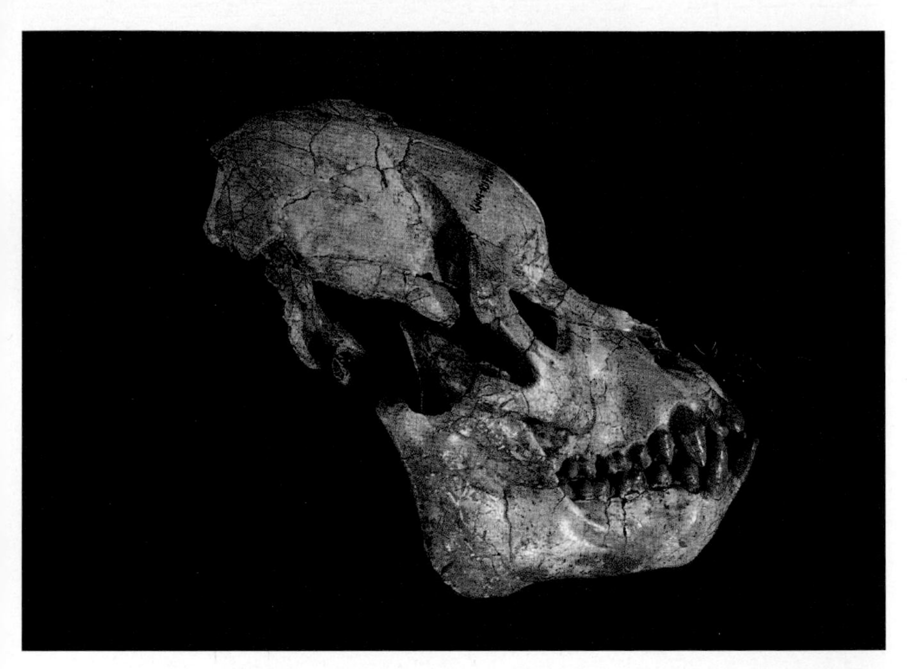

The original *Proconsul* skull was found in 1948. Additional pieces found in museum collections more than 30 years later glued on perfectly (right lateral view). (© Alan Walker.)

Two fossilized grass-hoppers (the one at the rear is upside down) that we found in a preserved foot-print of a deinothere near the R114 site on Rusinga Island. (© Alan Walker.)

Members of the expedition stand on Rusinga on the north flank of the Kisingiri volcano, looking across to the mainland, where the eroded center of the ancient volcano stands as an upraised plug. Faulting has lowered part of the former volcano, which is flooded by an arm of Lake Victoria. (© Alan Walker.)

The *Proconsul* forelimb skeleton and upper jaw—the original ape in a tree from site R114 on Rusinga—lie on a copy of the 1959 monograph on *Proconsul africanus* by John Napier and Peter Davis. (© Alan Walker.)

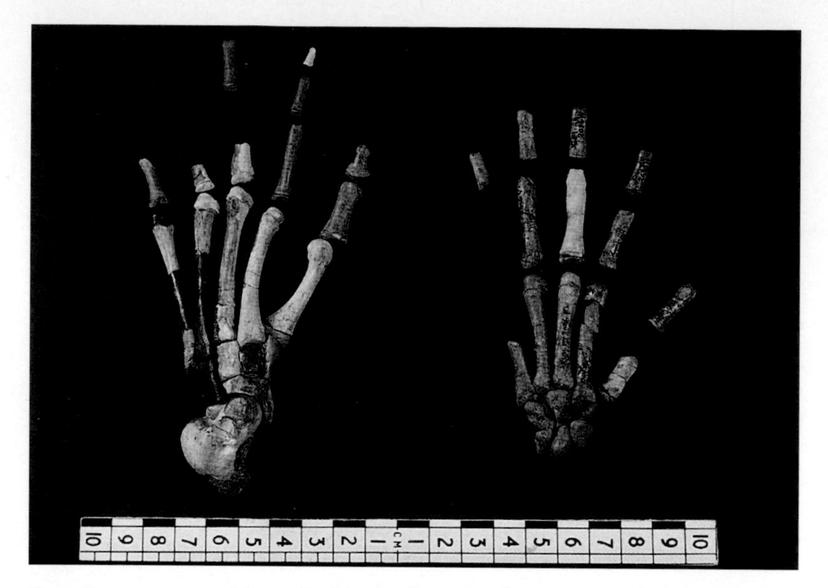

Once it was excavated from the block of matrix, the R114 foot skeleton could be reassembled and reunited with the original hand (right) and foot (left) bones that Napier and Davis had studied. Dark wax models are used for missing or incomplete elements. The varnish applied to the fossils excavated by Louis Leakey makes them medium brown; those excavated by me are covered in a clear modern preservative and look white. (© Alan Walker.)

When the troublesome generator was working, Maundu Mulila (right) and I (left) used airscribes to clean fossils in camp in 1984. Kamoya Kimeu stands behind me. (© Mark Teaford.)

The left and right halves of the upper jaw of *Proconsul* fit together again neatly although they had been separated for 34 years. One was found in 1950 and one in 1984. (© Alan Walker.)

Wambua Mangao (left) and Aila son of Derekitch (right) stand in Lake Victoria and wash sediment from the Kaswanga Primate Site through sieves. For once, the sieves were full of bones. (© Mark Teaford.)

We were astonished to find two adult *Proconsul* feet still articulated in the rock at the Kaswanga Primate Site. (© Mark Teaford.)

The camp at Kaswanga was nestled under a group of large fig trees that Louis and Mary Leakey camped under 35 years earlier. (© Mark Teaford.)

Iziah Nengo helped me excavate the first known *Proconsul* pelvis from the maize field site on Mfangano Island. (© Mark Teaford.)

The artist Jay Matternes did this reconstruction of *Proconsul heseloni* based on bones discovered on our expeditions. The image can be seen in a mural in the American Museum of Natural History, New York. (© 1991 Jay H. Matternes.)

3

An Arm and a Leg

Knowing that more fossils surely awaited them, Louis, Mary, and their colleagues continued to excavate at Rusinga and the other nearby Miocene sites until 1957. They collected thousands and thousands of specimens, but most of them were not primates; they were anything and everything but primates—mammals of all kinds and snakes, fish, mollusks and even fossil plants. There is even a fossil lizard with its scales intact and its tongue hanging out of its mouth and a small bird with flesh and feather impressions, both perfectly preserved in stone.

An extraordinary fossil site on Rusinga known as the fruit and nut bed yielded preserved wood, leaves, seeds, nuts, and fruits. The detail on the specimens from this site is amazing. Some of the fossil stems are wrapped around with creepers. Fossilized leaves still show the tunnels of leaf miners. Nuts and seed are preserved intact, looking—save for the loss of their original color—as if they had just fallen from the tree. There is even a fossilized fig with a bite mark on it, and I have often wondered if it was *Proconsul* who left that mark.

Most spectacular of all are the insects and other arthropods. A whole ants' nest was found and described by Edward O. Wilson, the ant expert from Harvard University. This was a hanging nest of leaves that must have dropped into carbonate mud and been preserved in-

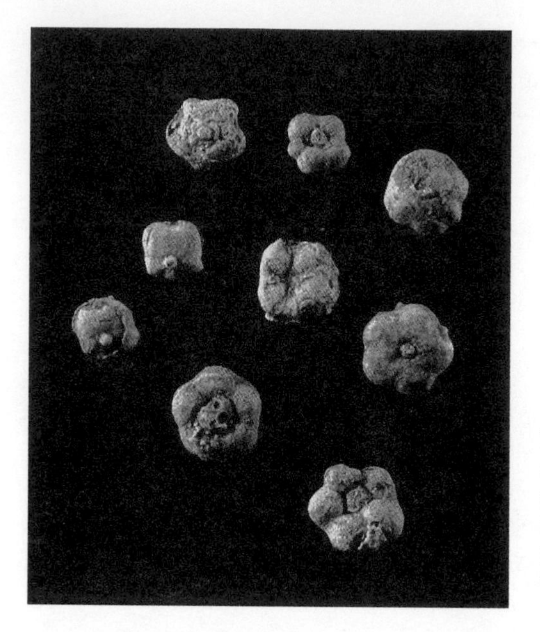

These nine fossilized fruits came from a single species of forest tree and were found at the Kaswanga Primate Site. (© Alan Walker.)

tact. Through careful preparation, Ed was able to extract ants and deduce the social organization of this extinct species from the size distribution of the workers. Though the species is extinct, the same genus of ant is still alive today in tropical rain forests. In addition to ants, there are fossilized spiders, grasshoppers, whipscorpions, assassin bugs, beetles—one bearing toothmarks from an insectivore on its elytra, or wing covers—caterpillars, pupae, insect eggs, and more. The grasshoppers were not represented by mineralized bones, as the mammals were, because insects have no bones. Some of these creatures died in volcanic soils that hardened around their bodies and others died by falling into wet, alkaline, volcanic ash. The soil or ash made a perfect, detailed mold of each one's shape. After their bodies rotted away, the empty cavities left in the rock were filled with calcium carbonate, which precipitated out of the groundwater and made beautiful geodes inside the cavities. The outer surface of these geodes reproduces the external surface of the grasshopper, assassin bug, or beetle, but when the geode is broken, there is no preservation of the internal structure of the original insect. These finds provide direct and irrefutable evidence of the existence of tropical Miocene forests in the area and of many of the creatures that lived in those forests.

To understand why Rusinga has such astonishing and unusual preservation, you have to know about the volcano known as Kisingiri. It is known as a natrocarbonatite volcano for the unusual combination of alkaline minerals, sodium, and calcium carbonate that it emits as ash and lava. There is a series of natrocarbonatite volcanoes along the eastern side of Africa, located in or on the shoulders of the Great Rift Valley, which were active at different times in the past. The only one of them that is active now is Ol Doinyo Lengai in northern Tanzania. The alkaline minerals dissolve in water rather quickly, making the groundwater very alkaline and the local soils very porous. Because the soils do not retain water, a sort of mock aridity is created despite the high rainfall. Thus the fossil soils at Rusinga indicate aridity, which at first seems to contradict the evidence of the fossilized plants, mammals, and snails, which are typical of rain forests. But this discrepancy is only apparent, an artifact caused by the unusual chemistry of Kisingiri and the mock aridity in the soils.

The prodigious amounts of volcanic ash and grit spewed out by Kisingiri during the Miocene accounts for the exquisite preservation of fossils on Rusinga and Mfangano. Trace fossils—impressions of the marks made by long-dead creatures—are also common. For example, when my expeditions went to Rusinga, we often saw impressions of the egg cases of beetles and we found a fossilized burrow with the skeleton of a giant springhare still inside.

When I looked south from Rusinga toward the mainland, I could see the heights of the southern flanks, which have not yet eroded away; they ring a pure limestone hill that sits in the middle. It is the remnant of a cold, alkaline lava that plugged the eruption vent of Kisingiri millions of years ago. After the volcano was no longer active, the northern flank eroded, Lake Victoria was formed, and Rusinga became an island.

Another stunning set of fossil finds was made by Thomas Whitworth in 1951. A geologist, Whitworth had been asked by Louis to map Rusinga to get a better idea of the chronological relationships among the fossiliferous beds. North of Kiakanga hill at the southwest end of Rusinga, Whitworth came across an unusual deposit of greenish, coarse, volcanic rock, or tuff. On the surface, the green rock was confined to a roughly circular area though the normal or country rock was gray; a little excavation showed that this distinctive green deposit continued vertically into the ground as a sort of pipe or tube-shaped deposit roughly 1 meter (about a yard) in diameter. Though the structure itself was intriguing, it was the contents of the deposit that were most fascinating: lots and lots of fossilized bones, many apparently associated into complete or at least partial skeletons. A later tally revealed that a total of thirty-one individual mammalian skeletons and five nonmammalian skeletons were collected from this one confined area during the 1951 season.

"The profusion of articulated skeletons found in this limited deposit suggests that it may represent the infilling of a pothole, in which animals were trapped," Whitworth said. The deposit soon became known as Whitworth's pothole, or the pothole site.

Among the finds from the pothole were the remains of a fossil pig, which Louis sent some years later to a Ph.D. candidate in England who wanted to study fossil pigs. There was also a nearly complete *Proconsul* forelimb, a few fragments and small bones from the hind limb of the same individual, as well as half of its palate and both sides of its lower jaw, teeth, and some skull bones. The bones of the forelimb were still articulated—preserved in their life position, or nearly so—and offered a tremendous opportunity to study the adaptations of *Proconsul*. There could be no doubt that the specimens all came from one individual animal; not only were they found in close proximity and in anatomical position, but the teeth and the limb bones independently showed this individual *Proconsul* to have been a juvenile animal at a particular stage of maturity. Of course, Louis sent the *Proconsul* bones to Le Gros for analysis.

At that time, Le Gros was involved in a particularly nasty and distasteful academic quarrel with a man who had once been a colleague of his in the Department of Anatomy at Oxford, which Le Gros chaired. The history between Le Gros and Solly Zuckerman, when the latter was a very junior colleague, was probably partly a motivating factor behind the dispute. As Le Gros wrote in his autobiography, "Within a few months of taking up my appointment [at Oxford] I was able to add to my teaching staff two junior demonstrators. One of these was Dr. (later Sir Solly) Zuckerman who was of considerable help to me in reorganizing and replanning of my Department." If you read between the lines, this description sounds as if Zuckerman, though in a subordinate position, immediately started trying to run the department. Zuckerman was a ruthlessly ambitious and relentlessly energetic young South African anatomist researching the endocrine regulation of fertility cycles in baboons. Even his biographer, a friend of Zuckerman's, admits that he was a man of "overbearing" and "autocratic manner" given to a "clearly-expressed conviction [that] he was right"—a man who carried out "merciless attacks on those of whom he disapproved, especially those he considered to be fools." To say Zuckerman must have been a difficult colleague for a soft-spoken soul like Le Gros to manage is probably an understatement.

In 1943, Zuckerman was offered the chair of anatomy at the University of Birmingham, after a lengthy wooing by Birmingham and much indecision by Zuckerman over accepting it. To be the professor of anatomy at Birmingham was definitely a step up; the problem was that Zuckerman yearned to hold the chair of anatomy at Oxford, a post Le Gros already occupied. In 1944, Le Gros advised Zuckerman that if he returned to Oxford rather than going to Birmingham, he could not expect to occupy as much laboratory space as formerly. Zuckerman went to Birmingham, resenting his treatment by Le Gros.

In 1950–51, Le Gros found himself sharing with Dart and Broom the dubious honor of being the target for a series of harsh papers emanating from Zuckerman concerning the ape-man, or australopithecine, fossils from South Africa. Zuckerman's main point, and the criticism that apparently unnerved Le Gros the most, was methodological. He suggested that the study of the shape of teeth or bones alone—the comparative morphological analysis in which Le Gros was so skilled—was old-fashioned and totally insufficient to determine the relationships among modern and fossil specimens. Put bluntly, Zuckerman charged that the way Le Gros had worked throughout his entire illustrious career was worthless. Zuckerman insisted that biometrical studies, based on repeatable measurements and the statistical analysis of those measurements, were essential. He roundly criticized scholars working in human paleontology who rarely showed "any understanding of the discipline of quantitative biology" and who indulged in "a wholesale neglect of modern statistical methods." His barbs were clearly aimed at Le Gros and the academic community knew it.

Writing with his student Eric Ashton, Zuckerman argued that Le Gros Clark, Broom, Dart, and others had made a fundamental error in claiming that the australopithecines were human ancestors. "Hardly one of the [australopithecine] teeth considered in this paper cannot be matched in dimensions and shape by the corresponding tooth of at least one type of extant great ape." Zuckerman disagreed with Le Gros's assessment that australopithecines were hominids and felt that they were more ape than man, as anyone with quantitative skills could have shown.

The replies and counter-replies flew fast and furious. Several morphologists pointed out that Zuckerman's techniques were hardly sophisticated; they were based on simple measurements such as length, breadth, and height of teeth. Indeed, such measurements were inadequate to distinguish between a sphere and a cube, much less between a gorilla and a human. The climax came in 1951, when two statisticians pointed out in *Nature* that Zuckerman's own statistical analysis of the dimensions of australopithecine teeth included a fatal mathematical error that, when corrected, negated his conclusions.

Zuckerman looked rather foolish, but the ferocity of his attack made Le Gros nervous about undertaking another analysis. The description of the *Proconsul* forelimb bones was bound to draw a lot of attention and the wise thing to do would be to incorporate some of these newer statistical methods. Sadly, Le Gros was never mathematically adept or statistically knowledgeable and he felt a new age was dawning in which skills he did not possess would be required.

The obvious answer was to pass the fossils on to a younger man who was more comfortable with statistical methods. Le Gros happened upon the perfect person to take over the task: John Napier, who would later become my graduate school supervisor and mentor.

John was a physician who had spent his war years repairing the hand injuries of wounded soldiers and trying to restore them to full function. In doing that, he became intimately familiar with the anatomy and function of the hand, learning how the hand worked to produce both subtle and strong movements. To enhance his understanding, he began studying the hands of other primates, too. After the war, John quickly became one of the leaders developing a new anatomical approach that linked form and function—an old anatomical theme—with evolution and adaptation. The comparative anatomy of living and fossil primates, and the ways in which their hand and foot anatomy reflected their habits, became John's lifelong passion. In the lecture room of the Unit of Primatology and Human Evolution that John created at the Royal Free Hospital of Medicine in London, he posted a large sign reading "Primatus sum, nihil primatum mihi alienum puto"—"I am a primate; therefore nothing about primates is irrelevant to me." The motto typified his eclectic, exciting approach.

John pioneered all kinds of research that focused on the biology

John Russell Napier, my mentor and graduate school adviser. (© J. R. Napier archives, University College, London.)

and behavior of extinct primates, rather than simply on their descriptive anatomy or their evolutionary position. He realized that locomotion, or the way an animal moves about the world, has a major influence on the size, shape, proportions, and anatomical specializations seen in the hands and feet. Those became subjects of greatest interest to him and I owe much of my own expertise in paleobiology and locomotion to John's influence and inspiration. I also owe to John my personal introduction to the *Proconsul* bones and the wonderful puzzles they have posed. Had he not delivered a brilliant and inspiring lecture at the Anatomical Society of Great Britain, which Le

Gros happened to hear, he might never have been given the job of analyzing the pothole *Proconsul.* And if John had not been asked to analyze *Proconsul,* perhaps my life and research would never have been so closely intertwined with this enigmatic ape.

In those days, there was so little documentation of primate behavior in natural settings that John had to amass his own data and develop his own techniques. He began collecting film footage from wildlife photographers, researchers at zoos, the BBC, and virtually anyone he could find with useful information. Slowly he began to categorize the habitual modes of primate movements into carefully defined locomotor categories: *arboreal quadrupedalism,* or moving in the trees on all fours; *terrestrial quadrupedalism,* or moving on all fours on the ground; *brachiation,* "the use of the forelimbs for grasping hand-holds above the head and swinging the body forward" in arboreal progression; and *terrestrial bipedalism,* moving on the ground propelled only by the legs and feet. (Some years later, I spent many hours observing bushbabies and lemurs in the wild and quantifying their locomotor patterns. John and I coauthored a paper adding another locomotor category to this list—*vertical clinging and leaping*—which some of those primates use to move rapidly through the trees.) In the 1950s, these were rather gross categories that were neither as well defined nor as well understood as they are now, but, to use a locomotor metaphor, they were a great step forward at the time. Before you can refine categories, you must define categories.

Proconsul offered a perfect opportunity to investigate the origins of the locomotor patterns of anthropoid, or higher, primates. The Rusinga specimen including the forelimb, now known as KNM-RU 2036 for Kenya National Museum, Rusinga, specimen number 2036, was among the oldest and most complete skeletons of a hominoid forelimb then known. As the introduction to the monograph on this specimen said: "It is clear that these bones belong to one of the most significant periods of Primate evolution; a period when the generalised catarrhine [Old World monkey] stock was emerging from a prolonged phase of arboreal quadrupedalism with its limited opportunities for adaptive radiation and was entering upon a phase that provided a diversity of environmental opportunity leading ultimately to the four distinct patterns of locomotion among the Anthropoidea."

In other words, *Proconsul* could give us a glimpse at what was happening among the higher primates at the time when the more primitive, monkeylike ways of moving around in the trees were evolving into a wider array of locomotor modes. New and different ways of moving meant that new and different resources—such as foods, nesting places, safe havens, and so on—could be tapped. These in turn would spur an adaptive radiation of stem apes. What a marvelous challenge to tackle!

John asked a colleague in the anatomy department at the Royal Free Hospital, Peter Davis, to join him in the project. They outlined an explicit research strategy. First they selected various modern species for comparison with the *Proconsul* fossils on the basis of their locomotor habits. Arboreal quadrupedalism was represented by a number of species that live mostly in the trees and move on all fours: *Cercopithecus*, the genus of many African forest monkeys such as Diana monkeys; *Pithecia*, the South American Saki monkey; *Cacajao*, the South America uakari monkey; and *Saimiri*, the South American squirrel monkey. Terrestrial quadrupedalism was represented by the baboon genus *Papio*, which moves mostly on the ground and spends relatively little time in trees, except when it is sleeping. All species in the genus *Papio* live in the Old World. Brachiation was represented by *Pan*, the chimpanzee; *Ateles*, the South American spider monkey, with its long arms and prehensile tail; and *Presbytis*, the leaf monkeys of Asia. Although now we know from field studies that adult chimpanzees do not spend as much time brachiating in the trees as, for example, spider monkeys and gibbons, this distinction was not clear when John and Peter were studying the *Proconsul* remains. The list of modern species that they used for their comparative sample of primates of known locomotor habits shows that they gave no particular preference to monkeylike or to apelike forms, to New World or to Old World animals.

A second criterion was used by John and Peter to select their comparative sample: immaturity. Various physical indicators showed that the pothole *Proconsul* had not yet reached physical maturity. In apes as in humans, during childhood the milk, or baby, teeth fall out, one by one, and are replaced by adult teeth. Since the last molars (the ones known as wisdom teeth in humans) had not yet erupted in the jaws of the pothole *Proconsul,* John and Peter knew the animal was not yet fully mature when it died. A similar immaturity was obvious in the limb bones of the animal, too. Although few humans notice the process unless they have cause to have an x-ray as a child, each long bone in the arms and legs arises from at least three separate centers of ossification, or bone formation: a shaft, in the center of the bone, and at least one epiphysis at each end. From birth until skeletal maturity, these centers of ossification are connected to one another by cartilage, not bone. Over time, the cartilage growth plates are replaced by adult, mineralized bone that then fuses the epiphyses to the shaft. The sequence of fusion throughout the skeleton is regular and predictable.

When John saw that nearly all of the epiphyses of the forelimb skeleton of *Proconsul* were still separate from the shafts, he knew the animal was a late juvenile. He and Peter were careful to use modern skeletons from animals in a similar stage of growth for comparison.

The final criterion was that they wanted to use skeletons of animals that had been shot in the wild, rather than zoo or circus specimens that might have had abnormal patterns of locomotion owing to confinement.

Then they began to take measurements of the fossil and modern bones and from them to construct indices or ratios of various measurements to discover which would be useful in distinguishing the animals of one locomotor category from those of another. They recorded means (averages) and ranges (size of smallest individual to size of biggest) of various measurements and examined the distribution of these figures by species and by locomotor category. They looked at the robustness of the bones and examined cross-sections through the various bones at defined points. Then they applied the same techniques to describe the fossil species' morphology and assess its locomotion. Nobody had done anything like that before.

After an enormous amount of work and measurement, they published a monograph in 1959 concluding that the upper limb of *Proconsul* "shows many primitive [monkeylike] and generalised features that provide evidence of a quadrupedal arboreal heritage; associated with these characters are others that appear to be adaptive for a brachiating mode of life. Specialisations of the fore-limb bones towards a terrestrial and cursorial [ground-running] life are entirely absent." The arm showed a combination or mosaic of features, suggesting the beginning of an evolutionary transition from arboreal quadrupedalism to a specialized use of the forelimb in brachiation. The analysis of the hand—which was complete save for a few small bones the dimensions of which could be reasonably estimated—added complexity by reiterating this mosaic pattern. As is common in monkeys, the hand itself was relatively short compared to the length of the whole upper limb. To have relatively short hands is the primitive or generalized condition, in contrast to the elongated palm and fingers of committed brachiators like the gibbon or spider monkey. Other hand features were clearly adaptive for brachiation.

Satisfyingly, an analysis of the plant remains from Rusinga published in 1957 by the paleobotanist Katherine Chesters indicated that the ancient environment at Rusinga was one of dense gallery forest festooned with lianas: a perfect place for an animal that combined adaptations for moving in trees on all fours with those for hanging from limbs and vines by its hands.

However, the issues of time and evolution had to be factored into this assessment. As John and Peter said in the monograph,

> The importance of the conclusion that *P. africanus* was a quadrupedal-brachiating form with a pongid type of dentition rests entirely on the interpretation of the term brachiation. If the onset of this mode of locomotion is regarded as having

> been an occurrence unique in evolution—a radical change in direction involving far-reaching consequences—then this fossil Primate may justifiably be regarded as the long anticipated annectant form between the quadrupedal monkeys and brachiating apes. On the other hand if, as we believe, brachiation is looked upon merely as an inevitable oft-repeated outcome of arboreal life, then the recovery from the East Africa Miocene of limb-bones demonstrating both quadrupedal and brachiating features should occasion little surprise.

They cautioned that the features of the limb bones were of little value for determining the placement of *Proconsul* within the primate family tree because they reflect adaptations to an arboreal way of life. Functional characteristics might arise many times in parallel owing to similar lifestyles, rather than being inherited within a single, closely related lineage. They agreed with Le Gros and Louis that *Proconsul* was surely some sort of primitive ape but did not feel they could make any firm pronouncements about whether *Proconsul* was the direct ancestor of any modern species and, if it was, which modern species that might be.

Louis, of course, still hoped and believed that *Proconsul africanus* was the direct ancestor of the human lineage. He displayed a tendency—common among fossil-hunters, I might add—to regard his own fossils as more unusual and important than anyone else's. As more bits and pieces of *Proconsul* were found in Kenya and Uganda by various researchers, Louis's attitude became more obvious. For example, I was amused to find a letter Le Gros wrote to David Allbrook in 1962, answering questions from David about possible *Proconsul* fossils that he and the geologist William Bishop had found at Moroto, a Miocene site in Uganda. Le Gros wrote:

> I have no doubt that your diagnosis of *Proconsul major* is the correct one. Don't allow yourself to be influenced by Louis Leakey—for some reason he seems to feel that every fossil *he* finds must be different from anything else, &, conversely, every

> fossil found by someone else must be different from his! He makes generic distinctions on the basis of quite trivial variations that may in some cases be no more than *individual* variations with the same species. He does not seem to be aware of the wide range of variability in the dental, skull and skeletal characters of anthropoid apes which has so often been stressed.

This letter was written shortly after I arrived at the Unit of Primatology and Human Evolution as a graduate student. By then, John had built a group so renowned that anyone interested in the field stopped by to visit, if he didn't actually join the unit. It was for me an extraordinary education, an exposure to an intellectual climate that was so brilliant and exciting that I woke up every day eager to get to the department and learn something new. The place fairly buzzed with new ideas. One of my assignments, shortly after my arrival, was to pack up the precious KNM-RU 2036 fossils so they could be safely hand-carried back to Nairobi, which provided a perfect excuse to examine them closely and compare them to the words and facts in the monograph. Just to be at a place where the analysis of these specimens had occurred, and where the techniques for deducing behavior from bones was being developed, made me feel that I was at the forefront of my science.

Three years later with my Ph.D. research on the locomotor behavior of living and fossil lemurs under way, I left London to take a job teaching anatomy at Makerere University in Kampala, Uganda, where David Allbrook was. I stayed in East Africa for the next 8 years, teaching anatomy first at Makerere University and then at the new University of Nairobi Medical School in Kenya. The medical students were good, my colleagues were excellent, and the work was worthwhile, but what I really loved about East Africa was the rare and exciting opportunity to do ground-breaking research. Science seemed wide open in those days; there was so much yet to do and learn.

The move to Africa took me closer to Madagascar, the only place where there are living lemurs. By sheer luck I met a wonderful and

The staff of the Anthropology Department of the British Museum (Natural History) in the early 1960s. John Napier is seated in the right foreground with his hands around his knee. I am the young and rather fierce-looking man standing with arms folded, in the back row. (© The Natural History Museum, London; Anthropology staff circa 1963, NHM Archives 178/14.)

slightly crazy character named Ike Russell. He was an American backcountry pilot who out of pure generosity made a hobby of flying researchers to remote places, free of charge. Because of Ike, I was able to fly to Madagascar to observe lemurs in the wild and to excavate at some of the sites in Madagascar where the not-yet-fossilized bones of recently extinct lemurs could be found. Everything I saw was new and seemed important because there had been so little scientific work done in Madagascar or Africa at that time. Even when the chance of a lifetime, such as flying to Madagascar with Ike, didn't come up, primates and other fascinating species were abundant in Uganda and Kenya. I fitted up a secondhand Volkswagen bus (known as a *kombi*) as a camper that would sleep my first wife and my small son and me. On weekends and vacations I drove to game parks and other wild places to watch animals in the wild and learn about their behavior and locomotion.

I became close friends with an extraordinary man, Jonathan Kingdon, a superb artist and wildlife ecologist who had grown up in Africa as the son of a district commissioner. Jonathan and I were close in age and shared a sense of adventure, a love of animals, and an aesthetic sensibility—I had very nearly gone to the Slade School of Art to study sculpture, rather than Cambridge to study paleontology. I was trying to quantify the locomotor behaviors I observed so that the crude categories John had defined could be turned into actual measurements. Jonathan had decided, in a moment of madness, to compile an atlas of all of the mammals in East Africa, incorporating original drawings of each species in life and from dissections, as well as maps of its distribution, and a summary of all that was known about its ecology and behavior. The task took him many years and the resulting volumes are exceptional for their breadth of coverage and the remarkable beauty of their illustrations. Together, Jonathan and I sometimes captured animals we wanted to study or bought them from locals. The ones that were not destined for dissection I kept briefly in large enclosures in my back garden, where I could observe them over a period of days or weeks before releasing them. While having a marvelous time with a congenial companion, I also learned a great deal about primate behavior, ecology, and anatomy that has stood me in good stead over the years.

During my years in East Africa, I also excavated some Miocene sites in Uganda, hoping to find more *Proconsul*s because I had had a special interest in the species since my graduate student days. The team of friends and colleagues I put together were rewarded with good field seasons in which we found many fossils of little-known and sometimes brand new species, but few *Proconsuls*.

Usually it was sites excavated by other people that yielded additional specimens of *Proconsul* in those years, but the most crucial discoveries related to the geology and dating of the sites. In 1967, the Americans John and Judy Van Couvering (who uses her maiden name, Harris, since their divorce) spent six months on Rusinga working out the age of the beds and the stratigraphic correlations among them. The geologic layer known as the Hiwegi Formation had yielded *Proconsul* specimens from the time of Louis's first excavations on Rusinga, so Louis recruited John to carry out his Ph.D. thesis at Cambridge and find out everything he could about the Hiwegi Formation. John collected the first samples from Rusinga to be subjected to radiometric dating, working on the project with Jack Miller of Cambridge University.

Until the 1960s, the only way to date relatively recent geologic beds was a sophisticated form of guessing. Logically, strata lying deeper in the earth were older than the layers overlying them. The hard slog of field mapping, of going out and walking along rock exposures and recording and measuring which ones overlay which other ones, has always been the primary basis of dating strata and the fossils in them. Then, in the 1960s, the potassium-argon method of radiometric dating was developed.

This technique is based on the principle that over time radioactive isotopes decay at a steady rate as they change to a stable form, known as a daughter isotope. One of the most useful radioactive isotopes in this context is potassium-40 (^{40}K) which decays to stable argon-40 (^{40}Ar). Volcanic eruptions produce intense heat that boils the rock and resets to zero the original ratio of solid ^{40}K to gaseous ^{40}Ar. The argon that has accumulated up to the point of the eruption is released into the atmosphere. Thus by measuring the ratio of ^{40}K to ^{40}Ar in volcanic rocks, we can discover how much decay has occurred since the eruption and arrive at a date for the eruption.

Using an early version of this technique, John and Jack dated the fossiliferous Hiwegi Formation to between 18.5 and 17.0 million years ago. This age was confirmed by a more comprehensive study made much later, using more refined techniques.

Like Rusinga, many of the Miocene sites in western Kenya that have yielded *Proconsul* fossils were dated a few decades ago. According to those findings, none of them is probably older than 20 million years. But I would love to see these sites redated now that a new version of potassium/argon dating offers improved precision.

This new method, known as single crystal laser fusion $^{40}Ar/^{39}Ar$ dating, can date very small samples, even single crystals of rock. Instead of determining the amount of potassium present in a sample, single crystal laser fusion dating uses nuclear technology to produce an artificial isotope, ^{39}Ar, that substitutes for ^{39}K, and then the ratio between the two forms of argon is measured.

If we use the older dates, the earliest *Proconsul* is from a site called Meswa Bridge. The site is said to be over 23 million years old but this date is very problematic. The authors of the original study that produced this date cautioned that it was inaccurate because their samples yielded two different ages—a problem that could now be resolved. Unfortunately, these two dates were mislabeled on a diagram, owing to a silly slip of the draftsman's pen, no doubt. That error made the dates appear to be consistent with each other at about 23.5 million years ago. This false date was then reproduced in later publications.

My best guess until more dating is done is that the oldest *Proconsul* is probably about 20 million years old. But the date by John and Jack of 18 million years for Rusinga was terrific. That made *Proconsul* a very old ape indeed.

4

The Lost and the Found

In 1980, a new era of *Proconsul* studies began accidentally, while I was visiting the National Museum of Kenya. (In 1972, the National Museum of Kenya in Nairobi was subsumed into a broader entity, the National Museums of Kenya, which includes regional museums. The "National Museum" still refers to the main museum in Nairobi.) My wife, Pat (coauthor of this book), was also there, working in the Palaeontology Department of the National Museum. One day as I wandered through Palaeontology looking for Pat to go to lunch, I saw that Meave Leakey—an erstwhile collaborator of mine and the wife of Louis's son, Richard—was unpacking some fossils. When I glanced over at them, the fossils were very familiar. Though they were obscured by a distinctive green sediment that I had never seen before, I immediately recognized the varnish that had been applied to the fossils, which had turned brown with age, and the awkward hand that had written R114 in large letters on them. The specimen in her hand looked just like the pothole *Proconsul* fossils that I remembered packing up as a graduate student, so I asked Meave if the fossils in the block of sediment were indeed from R114 on Rusinga. She confirmed that they were; they had been sent back by Albert Wilkinson, a pig expert in Britain who had had them for study. Then she unwrapped a piece of paper attached to one of the

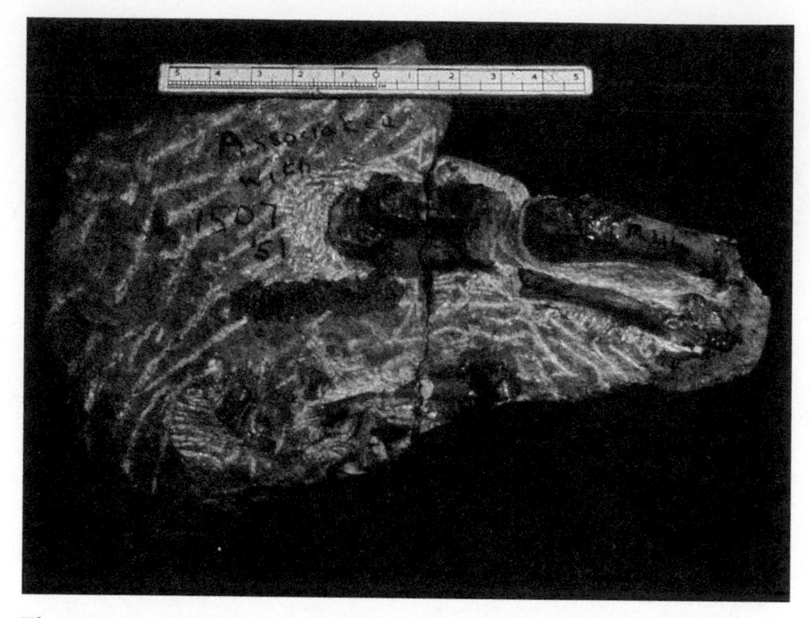

The unrecognized ankle and foot bones of the R114 *Proconsul* skeleton that were sent to England and years later returned to Kenya still in a block of rock labeled "Not a pig." (© Alan Walker.)

uncleaned lumps of greenish sediment, with bone sticking out of it, and read a terse message that made her laugh. She held it out to me and I read: "Not a pig."

Indeed, it was not. It didn't take Meave and me very long to realize that there was something in that lump of rock that looked suspiciously like an immature primate foot. After lunch, I took the lump off to the preparation lab to remove the rock. When I got the fossils extracted from the green matrix, I could see that we had found missing pieces of the 1951 pothole skeleton. There were large portions of the lower ends of both lower leg bones and a number of bones from the right foot. The new bones were in the same state of immaturity as the other bones of the pothole *Proconsul.* This new material fitted perfectly with the few foot bones that had been found by Whitworth in 1951 and described by John Napier and Peter Davis in 1959. What was arguably one of the most important fossil ape specimens ever had just become significantly more complete—and hence more important. I was elated. I'd always wanted to find a new *Proconsul.*

Suddenly a thought struck me: What if other specimens from R114 had been misidentified? I decided to locate every specimen in the museum that had ever been catalogued from R114 and have a good look at it. Martin Pickford, a geologist, was then working for the National Museums of Kenya surveying and gazetting prehistoric sites. Martin always liked working with fossils and since he was in Nairobi at the time, he was happy to help me on my treasure hunt. We tracked down every scrap of bone in the museum from R114 and laid them all out on a large table. The strategy worked beautifully. Among the misidentified or overlooked pieces we found new parts of the skeleton and many parts that fit perfectly with the old, previously known ones. In this case, the museum was the best place to look for specimens we agreed, grinning at each other.

The "excavating in a museum" aspect attracted some press attention, and before long reporters were asking questions like: "With all these experts around the museum, how did the specimens remain undiscovered for so long when they were already in the collections?" Only someone who has never participated in a dig that has yielded a lot of bones would ask such a question.

When an excavation yields only a few bones, those specimens are examined and reexamined exhaustively. But when a site yields thirty-one mammalian skeletons—and each complete skeleton has over 200 bones, a number increased further by fragmentation and by the five nonmammalian skeletons—some sort of triage must be performed. Whitworth made some preliminary field identifications of the bones as he collected them, but he was a geologist, not a paleontologist. Obviously Louis went through the fossils to decide what to send to whom, on the basis of his best guesses, but at the time the specimens were not yet cleaned of obscuring matrix. From 1951 until Martin and I reassembled the complete collection of R114 specimens in 1980, Louis was probably the only person who had ever looked at the entire assemblage—and he had made some mistakes. Lest someone think the experts to whom the bones were sent were careless, I would like to point out that the identification of broken specimens can be powerfully influenced by suggestion. If someone hands you a fragment of bone and tells you it is a pig, the image of "pig" is in your head and influences your interpretation. If you know pigs well and don't know much about, say, antelopes, then you might easily err with a fragmentary specimen. In this case, the pig expert knew some of what he had been sent was not a pig but he didn't much care what it was. And he didn't return the non-pigs to Nairobi until he was ready to return the pig specimens.

The new specimens offered the perfect chance to reevaluate the work John and Peter had done on the pothole *Proconsul.* Martin and I published a preliminary paper in 1983, which included a brief analysis and remarks on John and Peter's reconstructions of "missing" bones that were now found. One of these was the metacarpal of the thumb, a roughly cylindrical bone that underlies the fleshy, muscular part of the palm on the thumb side. Though the metacarpal was not present in the skeleton as John and Peter knew it, they had the bones that go on either side of the metacarpal and from those estimated its length. From the blocks I had cleaned, we now had a nearly complete metacarpal from the left hand and our estimate of its complete length was again a little under an inch long (25 millimeters): the same value John and Peter had estimated. When we found the complete right metacarpal in excavation at Rusinga, it proved that these estimates were within a millimeter or two of accurate. With the missing bones laid out with the others, I could assess the entire hand. My first impression was that the hand was rather primitive and monkeylike, though more detailed research would be required to prove this conclusion.

In 1959, John and Peter had estimated the total length of the humerus, or upper arm bone, based on about one-third of the bone. In 1981, we still didn't have a complete humerus but we had fully two-thirds of one. We estimated its total length to be about an inch (2.5

centimeters) shorter than they did, which of course made the entire arm shorter. Still, when we examined the proportions of the segments of the arm by using the brachial index—the length of the forearm divided by the length of the upper arm—we found *Proconsul*'s arm had the same proportions as the arm of a modern chimpanzee. This finding strongly suggested a brachiating habit in *Proconsul,* as John and Peter had concluded. The mobility of the shoulder and elbow joints also looked rather chimpanzee-like.

What we had, then, was an animal that was apelike at the shoulder and became progressively more primitive and monkeylike toward the hand. *Proconsul* was indeed an interesting primate.

The additional specimens of the hind limb offered different insights. The hind limb was surprisingly sturdy for an animal with a body weight of only about 24 pounds (11 kilograms). The best match in size and robustness that I could find in the osteology collection of the National Museum in Nairobi was a male black and white *Colobus* monkey, and male colobus monkeys may weigh up to 30 pounds (13.6 kilograms). The foot showed a big toe set well apart from the other toes, as in chimpanzees, that could be used for grasping branches. Farther up the leg toward the hip, the leg bones were more monkeylike. Thus the forelimb of *Proconsul* had a gradient from more apelike to more monkeylike, going from the shoulder to the hand, whereas in the hind limb this gradient is reversed, with the monkeylike anatomy at the hip and the more apelike features in the foot.

The overall image of *Proconsul* gained from all this information is of a powerful, fairly slow-moving animal that spent most of its time in the trees, moving both on all fours and also hanging from branches. In many ways, what we were able to do is confirm the preliminary conclusions reached by John and Peter in 1959 and elaborate them with more information and some new bony specimens.

Something else came out of these studies: a new idea. If Louis had misidentified the leg of *Proconsul* and a number of its arm bones in Nairobi, had anything been left behind at the site? I had long supposed that nothing was left of the site. Peter Andrews, a paleontologist and occasional collaborator of mine, had just written a short history of the fieldwork in Miocene fossil sites in Kenya. Peter had been to practically every one of the Miocene sites himself, since they were a special interest of his, and he had written, "The exact nature of the supposed pothole deposits [on Rusinga] remains in doubt because subsequent excavation completely removed all contextual evidence, but it may have been a trap in which entire animals were preserved." Now I wondered if that was precisely true. Was the pothole site gone—destroyed—or simply difficult to find?

I decided to go to the documentary sources first. Martin found Whitworth's field diary from 1951 in the National Archives of Kenya and read his notes on the R114:

> *August 27th, 1951.* Continued mapping Gumba peninsula and on western slopes of Kiakanga accidentally stumbled onto what seems to be the best find to date—a complete skull of a ??pig with an articulated carpus [hand] and various sundries.
>
> *August 28th, 1951.* Today Ngunjiri [a Kenyan assistant] commenced to dig out various odds and ends of skeleton from the pig site which I discovered yesterday, whilst his collectors searched the wash area below the outcrop. Ngunjiri seems determined to move about half a ton of boulders back to the Museum for preparation, maintaining stoutly that this is the official Coryndon [now the National Museum in Nairobi] technique.
>
> *August 29th, 1951.* Collecting party concentrated on the pig site. Found almost every fragment of rock in this tiny and isolated outcrop is crammed with bones, teeth etc. The chief find today was a mandible of ?*Proconsul africanus,* the dentition lacking only M3 on each side and the right canine. With it were a number of fragments of a cranial vault which may appertain. Whilst keeping an eye on these proceedings I continued my geological mapping.
>
> *September 4, 1951.* Collecting party in flaggy beds in immedi-

ate vicinity of agglomerate pipe worked during the last few days. This latter site Ngunjiri wishes to leave until he can secure more suitable equipment on some future visit—and no doubt more suitable and more skilled supervision.

Whitworth sent some blocks of rock back to Nairobi and then, in December 1951, Louis went up to see the site for himself. Louis's diary reads:

> *December 16th, 1951.* Went to Rusinga with Mary and the 3 children, Heselon and Ngunjiri. Made camp at bay S. of Gumba and went up to see the block found by Whitworth. Found it to be a filling of agglomerate in a pothole in the rock. Suggesting a break in deposition. The block is full of bone and we started off by "stone-breaking" the mass that Whitworth left in a pile. From 16th to 29th broke this and packed 18 boxes of material worth taking.

It was the last phrase that made me catch my breath: "18 boxes of material worth taking." Louis's words implied that he could have taken more than 18 boxes of blocks, but he didn't think some of the blocks contained good fossils and so left them behind. Given that both he and Whitworth described the bone in the rock as abundant, some of that rock left behind might contain bones I very much wanted to see.

Had important fossils been overlooked at the site and left behind? Could I find the site again by following Whitworth's notes and maps? I had to go see for myself.

I was eager to try to find the missing pothole site, but my teaching responsibilities in the United States forced me to postpone a visit to Rusinga for a while. Martin Pickford, the geologist, had caught the idea and in his spare time perused Louis's other field diaries from Rusinga, to see if there was anything else that warranted investigation. In Louis's 1947 diary, he found a tantalizing entry from Louis's time on Rusinga about finding some turtle scutes, which are the flat plates of bone that underlie a turtle's shell. The location was very near the site where, a year later, Mary found the beautiful little *Proconsul africanus* skull. Martin knew that turtle scutes had been mistaken for skull fragments before, and vice versa. He wondered if the "turtle scutes" were actually some of the missing pieces of Mary's skull that had been collected but had gone unrecognized, just like the pieces of the pothole *Proconsul.*

Martin decided to look at the original "turtle scute" specimens himself. Where exactly he found the specimens he never said; in fact, he was a little evasive on the subject. Though at first I assumed he'd found them in the unaccessioned and unidentified material in the National Museum of Kenya in Nairobi, in retrospect I don't think the fossils were in those collections in Nairobi because little or nothing from Louis's 1947 expedition was. Most of the material from that expedition had been sent to the British Museum (Natural History) in London. Wherever Martin found the box of fossil fragments from Rusinga, he looked through them and extracted two that looked like skull fragments rather than scutes. Then he waited until he could check whether or not they glued onto Mary's little skull.

Mary's skull had been in London since 1948. When Richard Leakey, Louis's son, became the director of the National Museum of Kenya in 1968, he began trying to persuade the British Museum (Natural History) to return the *Proconsul* skull to Kenya. Richard had been told by his father that the skull had been taken to Britain *on loan.* Ownership of the skull was a matter of national pride, since the fossil was one of the greatest finds ever made in Kenya, and as the new director, Richard was perhaps trying to establish his credentials. However, he found that, at some point, the *Proconsul* skull had been accessioned and registered by the British Museum (Natural History) under the number M32363. The M stands for fossil mammal, and the number indicates how many mammalian fossils had been accessioned before that specimen. Clearly the British Museum believed the specimen had been *donated* to it, not lent, and it was not inclined to part with one of its treasures just because someone asked for it.

The stalemate continued until June 1981, when Hazel Potgeiter, Mary Leakey's longtime secretary, came across a crucial letter to Louis Leakey in the National Archives of Kenya. The letter was from the acting chief secretary of the Kenyan government, whose signature is illegible, and was written on November 21, 1949. In it, the acting chief secretary discussed the conditions under which the skull "may be housed in London for the time being . . . [with] the clear understanding that the Kenya Government retains ownership and reserves the right at some future date to resume possession." Nothing could be clearer: the specimen had indeed been lent to the British Museum in London. Delighted, Richard sent off another formal request for return of the skull to the appropriate authority, Dr. R. H. Hedley, including a copy of the letter from the acting secretary.

I can imagine the consternation at the British Museum when Richard's new request and documentation arrived. Documents in the Natural History Museum's archives show that the trustees and public relations officers of the British Museum were aghast. The newfound letter showed that they had been wrong to accession the skull, which indubitably belonged to Kenya. Various officials at the British Museum worried about the best way to handle the matter. When—not if—the story came out in the press, the museum could look very bad, like a colonial institution that had looted and pillaged the Third World for its treasures. The simplest solution was to downplay everything by deaccessioning the fossil skull and returning it to Kenya as graciously and rapidly as possible, which they did.

Mary's *Proconsul* skull returned home to Kenya on October 25, 1981. The newspapers and magazines were full of headlines. "Ancient Skull Goes Home" one crowed; "Victory in Battle of Skull" another declared; and a third said simply, "The Return of the Skull." When the skull came back to Nairobi, Martin and I were able to compare the purported scutes with the broken edges of the skull. The scutes

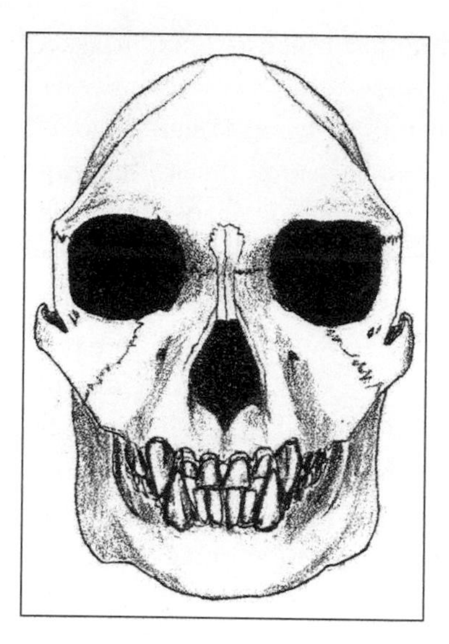

Once we found additional pieces of Mary's 1948 *Proconsul* skull, I drew this reconstruction (anterior view), which minimizes the distortion seen in the fossils. (© Alan Walker.)

fitted perfectly. The new pieces formed a solid, bony join between the main part of the skull and the base of the skull, or occiput, which had until then always been separate.

With the newly enhanced skull, and the more complete pothole skeleton, I could finally ask what sort of brain size *Proconsul* had for its body size.

I called in some colleagues to collaborate in the analysis. Dean Falk was a well-known paleoneurologist, now at Florida State University, who had a large database on the size of the body and the brain in living primates. She had examined the little *Proconsul* skull and a cast of the cranial cavity in London before the new pieces were added. At that time, she had said in print that she believed the skull was too incomplete for an accurate estimate of brain size to be made, even though some of her colleagues had published brain size estimates. (Sometimes Dean is a little sharp tongued and impatient with those less clever or less careful than she is.) Now, with a more complete skull, she was eager to see what its cranial capacity was. The other colleague was Richard Smith, who was trained as an orthodontist and was then at the University of Maryland Dental School. Teaching the future dentists of America did not engage Rich's active brain sufficiently, so, when he could get away, he sought intellectual stimulation by interacting with the functional anatomy group at the Johns Hopkins University School of Medicine. He certainly stimulated us, for Rich was an expert in the statistical techniques for measuring and comparing the variability in brain and body size across species. He has a thoughtful and clear-minded way of summing up problems that makes him an excellent collaborator. Together, we attacked the challenge of determining the brain and body size of *Proconsul.*

Though many pieces of the cranial vault were still missing, and the skull had been squashed obliquely while it was in the ground, I thought we had enough to get a good estimate of the cranial capacity of *Proconsul.* I took the occasion to do a little additional cleaning of matrix from the fossil and to make an accurate plaster cast of the entire skull with the new pieces. Wherever there seemed to have been distortion or folding of the bone while it was in situ, I cut the cast into pieces. Then I reassembled the many tiny fragments, following clues such as left-right symmetry, the need for the two halves of the lower jaw, or mandible, to meet at midline, and the need for the upper and lower teeth to occlude in such a way that the observable dental wear could actually have occurred during life. Rereconstructing Mary's little skull was exactly the sort of three-dimensional anatomical puzzle that I relish. When I was done, a good deal of the distortion had been removed from the specimen and it no longer appeared to have such a long and pronounced snout. A peculiar depression of the nasal bones had also disappeared.

Unfortunately, the skull was still too incomplete to measure the volume of its cranial capacity directly by pouring in water or seeds and then measuring the volume that filled the cranium. To make the cranium complete enough to hold any substance, wet or dry, I would have had to fill in a great many gaps with Plasticine, which always involves a tremendous possibility of being wrong. What we could (and did) do was measure the length of the arc (the curved interior surface) of the braincase along the midline from the front to the back. This arc was defined by solid bony connections all the way along, with minimal distortion. By taking similar measurements of modern primate skulls of known cranial capacity, we were able to prove that the length of the arc is so closely related to brain volume that one measurement can be used to predict the other, using a fairly simple mathematical equation. We plugged in the value of the arc length from the fossil and estimated the brain volume of Mary's *Proconsul* skull at 167.3 cubic centimeters.

Now all we needed was a good estimate of the body weight of *Proconsul africanus.* The body weight of an extinct species is often estimated by using the relationship between some linear measurement, preferably of a bone that regularly supports the body, and the known body weight of recent animals. For example, the diameter of the femur, or thigh bone, is a very good predictor of body weight, which makes sense since the femur must support the body weight. If we had a femur with a diameter that could be measured, we could estimate body weight—but Mary's skull had no associated limb bones. Fortunately, the pothole *Proconsul* was from the same species, came from a site not far from the one where Mary's skull had been found, and was of the same geological age. Both the skull and the skeleton appeared to be female. Assuming—because we had no choice—that both specimens were approximately average for their sex and species, we could use the pothole *Proconsul,* which had all the limb bones anyone could want, to estimate body weight for the species. We could have estimated body weight many times over from the length or diameter of each limb bone, which would have produced a confusing welter of values. What we did instead was rely on the simple method of matching the overall size and robustness of the *Proconsul* skeleton with skeletons of living species of known body size. Once again, the closest matches we could find were males of the black and white colobus monkey, *Colobus guereza,* which had average body weights of about 23 pounds (10–11 kilograms). Being conservative, we estimated that

P. africanus females had a body weight of 22.2–24.2 pounds (10–11 kilograms) to go with the brain size of 167.3 cubic centimeters. What did these facts mean and how should we interpret them?

Comparing relative brain and body sizes among species is a thorny matter. One common technique is to calculate an Encephalization Quotient, or EQ, which ignores the absolute value of brain or body size and instead expresses each species' ratio of brain to body size as a percentage of the human ratio. Thus a hypothetical animal that is as relatively brainy as a human has an EQ of 100 percent, while a theoretical animal that is literally all brawn and no brain has an EQ of 0 percent.

According to our estimates, *Proconsul africanus* had an EQ of 48.8 percent. That fact alone wouldn't tell us if *Proconsul* was an ape or a monkey, because living monkeys and apes have overlapping ranges of EQs. Monkeys' EQs range from 22.9 percent to 82 percent and ape EQs range from 17.2 percent to 41.1 percent. Thus the EQ of *P. africanus* was larger than that of any living ape but wasn't outside the range of monkey EQs. And, as I've said before, the real question about *Proconsul* isn't whether it was a (modern) monkey or a (modern) ape, because it wasn't like either of them. The issue for me is what *Proconsul* was like in and of itself and which features of apes had already evolved at this early point on the ape-human lineage.

The EQ wasn't going to tell us if the fossil species was smarter or dumber than moderns apes or monkeys, because neither EQ nor absolute brain size has any known relationship to anything that could be measured as intelligence. Measuring intelligence in humans is difficult enough, as thousands of pages in learned books and articles attest. Measuring intelligence in monkeys or apes must be a quagmire of complications.

It was tempting to compare the EQ of *Proconsul* with that of the living great apes. Rich advised caution, reminding us that comparing the EQs of species that differ considerably in size is often misleading even though an EQ is supposed to compensate for absolute body size differences by measuring the amount of brain per unit of body size.

Nonetheless, animals with bigger *absolute* body sizes tend to have lower *relative* brain sizes, hence lower EQs. For example, gorillas have the highest body size and the lowest EQ of any ape, at 17.2 percent; orangutans are intermediate in value; and common chimpanzees have the highest EQ. We couldn't be sure what, exactly, was the meaning of the fact that *Proconsul africanus* had a higher EQ (48.8 percent) than any of the living great apes since all three living species are so much larger in body weight.

At 25–30 pounds, *Proconsul* was similar in body size to some of the lesser apes, such as the gibbons of Asia, and had a higher EQ (48.8 percent versus 41.1 percent) than the living lesser apes, for what it was worth. The question of diet complicates the interpretation of these facts, because gibbons eat varying amounts of leaves and fruits and leaf eaters tend to have smaller EQs. Such tangled complications plagued our research on *Proconsul.*

The only comparison that seemed relatively pure and simple was between the EQ of *Proconsul* and that of various monkey species of the same general size. With Dean's help, we were able to find good data on brain size and body size from eleven different species so that we could calculate their EQs. This monkey sample had EQs ranging from 22.9 percent to 40.9 percent, with an average of 30 percent. With its EQ of 48.8 percent, *Proconsul africanus* was distinctly larger in relative brain size—more encephalized—than any of these monkeys, as well as being more encephalized than any of the living apes.

This result had powerful implications. *Proconsul africanus,* from all we knew of it so far, was a primitive hominoid, or stem ape, dated to 18 million years ago, yet *Proconsul* had a degree of encephalization bigger than that of modern apes and monkeys of similar size. The conclusion I drew was that apes and monkeys as a group have not gotten much larger in relative brain size during the last 18 million years. The big brain often taken as a hallmark of higher primates is very old indeed.

Could our estimates be faulty? Just to test our certainty, we asked ourselves how wrong we would have to be to make the EQ of *Procon-*

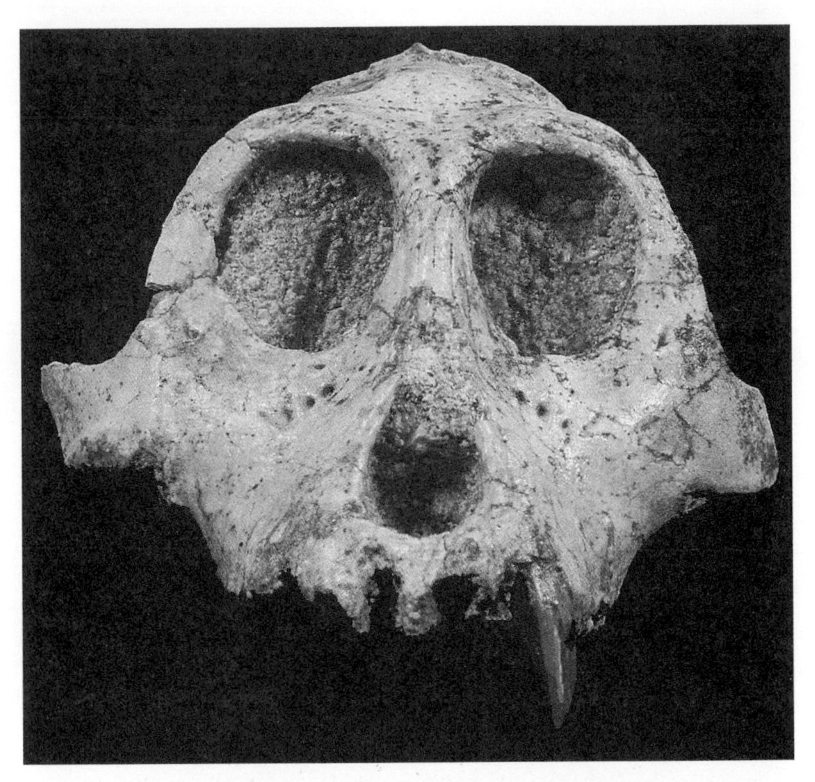

This *Victoriapithecus* specimen is the oldest known monkey cranium, 15 million years old. It was discovered at Maboko, Kenya. (© Brenda Benefit and Monte McCrossin.)

sul equal to that of a typical monkey (30 percent) in our sample. To produce an EQ of 30 percent, *Proconsul*'s body weight would have to be almost twice as large as our estimate, or just over 47 pounds (21.4 kilograms). None of us believed that our estimate was so very far wrong. This exercise in "how wrong could we be?" made us confident enough to assert that *Proconsul africanus,* 18 million years ago, had an EQ comparable to that of modern monkeys and apes. This hinted that something fundamental about being an early stem ape (and not an early monkey) might have to do with relative braininess.

In 1997, the discovery of an Old World monkey skull that was 15 million years old was announced. The species *Victoriapithecus,* to which the skull belongs, is moderately well known from other specimens, many from Maboko Island, Kenya, where the skull was found. By using the postcrania of the same species to calculate body weight, the EQ of the new skull was determined to be 34 percent: about the EQ of a *modern,* small-bodied monkey, and yet the *Victoriapithecus* specimen is the oldest Old World monkey skull known at present. Only these two specimens—one of the ape *Proconsul* and one of the monkey *Victoriapithecus*—offer solid evidence of EQs of higher primates from the African Miocene. Although the EQ of *Proconsul* is substantially higher than that of *Victoriapithecus,* together these skulls suggest that modern levels of encephalization have a very ancient history among both monkey and ape lineages.

In 1982 I went to Rusinga with Martin to look for the pothole site. Finding the highest point on Kiakanga Hill was not hard and Whitworth had recorded a precise location for the pothole from that point, based on a compass bearing and a linear measurement. It was like following a map to a buried pirate treasure. I held the compass and gave hand signals to move left or right, while Martin paced off the distance. When he finally stopped, he looked around and yelled back, discouraged, "There's nothing here!"

"Oh yes there is!" I answered gleefully. What he couldn't see from his position, but I could from a greater distance, was that he was standing in the middle of a roughly circular patch of bright green vegetation in the middle of a dry, straw-colored plain. *He was standing on the pothole.* As soon as I pointed this fact out, Martin saw it too. We scrabbled around a little, found a few fragments of bone, and located Louis's "dumps"—the pile of material "not worth taking," sometimes called backdirt, that virtually any excavation creates. We had found it.

5

Back to the Miocene

When I got back home to the United States, I wrote proposals to a few foundations for a grant to reopen excavations at Rusinga. By early 1984, I had secured a National Science Foundation grant to explore Rusinga and revisit the pothole site, with collecting additional fossils and resolving the exact nature of the pothole as major aims of the project. I put together a team drawn from the National Museums of Kenya and the Johns Hopkins University School of Medicine, where I was then employed. John Ndere from Rusinga, who was a sort of caretaker for the sites on Rusinga, also helped us for a time.

At that time, Richard Leakey and I had been working together for years finding early hominoid fossils in Kenya and researching them, and we were due to start work soon on the west side of Lake Turkana. I decided to go to Rusinga before my field season with Richard.

When Richard was a child, he was dragged by his parents to Rusinga, where he was bored and miserable. Work started at dawn for Louis and Mary, and the children—Jonathan, Richard, and later Philip—were expected to entertain themselves. When Mary was excavating the *Proconsul* skull, Richard was only 4 years old. He still remembers watching his mother working doggedly in the heat. He hated sitting in the sun, swatting flies and waiting for a breeze that never came, but he did not want to leave his mother either. He desperately wanted her to pay attention to him and she was transfixed by fossils. On that occasion, he vowed with childish determination that he would never, ever, pursue a profession that required excavating in the blazing sun. When he revisited Rusinga in 1984 with me, he was well known as a highly successful paleoanthropologist—pursuing the very career he had sworn to avoid. The sight of Mary's tree immediately brought back his memory of sweaty misery and a childish vow that he had long forgotten and certainly broken. Though the work in northern Kenya was in an even hotter climate than that at Rusinga, Richard's dislike of the island persisted in his adult life. He assisted the expedition by flying supplies and visitors up from Nairobi, but Richard's heart was not in the Miocene work and he spent little time at the site. Perhaps his most valuable contribution was to allow me to take many of the best fossil-finders employed by the National Museums of Kenya, a team known as the Hominid Gang that he had built up over years.

The leader of the Hominid Gang was Kamoya Kimeu, an extraordinarily able and knowledgeable man that I count among my closest friends. I have trusted him with my life and would do it again. Though he never had much formal education, Kamoya is a natural manager of men and probably the best hominid fossil-finder in the world. Usually he looked for Plio-Pleistocene hominids up at Lake Turkana. Now I was taking him back in time—back to the Miocene 18 million years ago—and to a different lake to look for apes, but Kamoya is always happy if he and his "people" are finding fossils. At that time, his people included Maundu Mulila, Wambua Mangao, Musa Kyeva, Benson Kyongo, and Peter Nzube (usually called by his surname, Nzube), the core of the Hominid Gang. Like Kamoya, all of these men were from the Wakamba tribe and had been raised in Kamoya's home village near Machakos, about 40 miles southeast of Nairobi. Maundu was then the heir apparent of the Hominid Gang, the young and talented man with the right gifts and personality to take over when Kamoya retired—at least that's what we all expected.

Aila and Solomon sons of Derekitch were two brothers from the

In 1984, the Hominid Gang poses with me (back row, far right) and Mark Teaford (center front) at the R114 site on Rusinga Island. Middle row, left to right: Kamoya Kimeu, Solomon son of Derekitch, Peter Nzube, Musa Kyeva. Back row: Maundu Mulila, Wambua Mangao, Aila son of Derekitch, Benson Kyongo. (© Alan Walker.)

Dassenech tribe in the Turkana region. They came along on this expedition as they had on many others, Solomon serving as the camp cook and Aila working as a fossil-hunter. They looked different from Kamoya's Wakamba men and their culture was quite different, but they knew their jobs and did them well. Not only was the first language spoken by Aila and Solomon different from that of the others, but it was not even closely related to their native tongue, Kikamba. (Among Bantu peoples, generally the root word—in this case, "kamba"—is preceded by "Wa" to refer to the tribe and "Ki" to refer to the language.) Fortunately, everyone on the team spoke some Kiswahili, including me, although my command of Kiswahili is limited to the strictly functional phrases needed to run an expedition. I have only rudimentary Kiswahili and cannot even ask for news of the men's families except in English. Each of the men also spoke English with varying fluency. Martin Pickford came along that first season to revise his maps of the site; Martin and I spoke British English and Martin, who had been raised in Kenya, also spoke fluent Kiswahili. The last member of the team that year was my postdoctoral fellow, Mark Teaford, who was again very different from everyone else. Mark was from an American family that had never had a college graduate before him, much less a Ph.D. He had traveled little in his life before 1984. This was his first trip outside the United States, his first trip to Africa, and his first experience hunting fossils. He was completely agog. From time to time, he seemed a little overwhelmed by the novelty and the strangeness of everything but he tried hard to fit in and pull his weight. Somehow everyone got along well in this multicultural and multinational team, despite our differences.

We left Nairobi in two parties—one by lorry, the other by Land Rover—and met up on May 12 in Kisii, well upcountry. From there, we chugged in convoy to Mbita Point, where the causeway from the mainland to Rusinga starts. When we arrived, it was after dark on a Friday night, so it was too late to check in with the District Officer, or D.O. We went on to Rusinga to meet with the local chief, who graciously received us in his home. We camped that night at Kaswanga and set off early the next morning to pick up John Ndere and find a route through to Kiakanga. We found one, but it involved wending our way carefully through the small subsistence farms—*shambas*—that seemed to be everywhere on Rusinga.

We made our permanent camp one gully over from the "pothole" site, where the black kites flew overhead and played aerial games with their mates. By then, we were pretty doubtful that the site was a pothole and so shifted our terminology and called it R114 most of the time. The lake would be our water source, bathtub, and recreation facility as soon as we found a shorter way to get to the lakeshore from the camp. Until that detail was taken care of, we could obtain lake water from a tap at the Tom Mboya School nearby, but it had to be boiled for drinking or cooking. The kitchen was artfully situated under a tree that would provide lots of handy "hooks" for hanging up pots and pans, the dining tent was set up with a good view over the lake, and the toilet (which we called by the Swahili word, *choo,* which is pronounced to rhyme with "show") was a wooden seat mounted over a deep hole and sheltered for privacy with a canvas screen. We erected our sleeping tents, doled out the camp beds, mattresses, pillows, sheets, blankets, lanterns, and wash basins, and suddenly we had a pleasant, tidy, and comfortable home until the end of June. The gang were all thoroughly experienced in setting up camps like this and the process went smoothly.

That night, we experienced a special Rusinga event: a thundering, roaring, the-end-of-the-world-is-coming storm with lightning strikes all around us. No one could sleep, especially me, because my tent (last used by the geologist, I noted) turned out to have no waterproofing at all any more. The lack of waterproofing was exacerbated by several strategically placed holes that served as conduits for large quantities of rainwater. The next day, my pathetic, soggy tent was taken down and labeled for repair and another was erected; we set up fly sheets to help shelter the tents and dug trenches around each tent for drainage.

This first storm was the beginning of a nearly nightly light show accompanied by thunder so loud it was hard to talk over the noise. Inclement weather interrupted dinner often enough that we took to moving the *jikos*—iron charcoal grills—into the dining tent at the first hint of rain so the cook could keep the fires going.

Kamoya got the radio set up and called Nairobi to tell Richard we had arrived safely—staying in radio contact was one of his usual responsibilities on expeditions—but he forgot it was a Sunday and spoke to the museum watchman, who was the only one there. Never mind; at least we had checked in so no one would worry about us. The next morning the men and I sorted out the camp and built small stone walls to protect some fossils that were weathering out. Martin thought the fossils were from a tragulid (a small water deer, or chevrotain) but they proved to be a springhare instead. We had also heard some bones were coming out at Kaswanga and we went later in the day to examine them. Still mostly buried was a large jaw with huge canine teeth: not a primate, but interesting nonetheless. The people there also showed us a few *Proconsul* teeth that had been found loose on the surface. It was only our second day on Rusinga and we already had new *Proconsul* specimens.

On the third day, we settled into the work we had planned. I had brought with me from the States several devices known as airscribes. About the size of a ballpoint pen, an airscribe is the equivalent of a miniature jackhammer and it is perfect for cleaning fossils without damaging them. We had hauled a very expensive air compressor all the way from Nairobi to power the airscribes as well as two binocular microscopes that enabled whoever was cleaning fossils to see every tiny detail. This was, to my knowledge, the first time anyone had tried to set up a fossil-preparation laboratory complete with state-of-the-art instruments in the field. Our plan was to go through Louis's dumps and check every piece thoroughly for teeth or bones. The airscribes would let us clean the green matrix off anything that looked promising without damaging it. We had no intention of leaving pieces behind and having someone find out decades later that we had abandoned something important on the site.

Setting up the outdoor prep lab was more difficult than I had assumed it would be. The first problem, of many, was that I had forgotten to bring one particular connector that was needed to make the whole airscribe and air compressor apparatus work. We radioed Nairobi to get someone to bring the connector up. Of course, when it arrived two days later, the new one didn't fit, so we found a metalworker in the closest sizable town on the mainland, Mbita; he managed to braze it onto the air compressor's tap. Yet another day passed while he worked before we could start cleaning the fossils.

In the meantime, we set about making a huge heap of all the lumps of green rock from R114 that Louis had left behind. We examined each one. If no bone showed when someone picked it up, the lump was set aside. Later, a rotating team would work like convicts breaking rocks, hitting each lump with a hammer until either bone

Sitting in the heat haze with Lake Victoria in the background, three members of the Hominid Gang break rocks from Louis Leakey's 1950 dumps to find *Proconsul* bones. (© Alan Walker.)

was revealed or the remaining piece was too small to contain anything. Clearly, many of the lumps were full of bone and we were elated, itching to get the airscribes up and running so we could see what lay within. Wambua found an especially promising piece that showed a fragment of tooth and brought it over to me. I looked at it and said, "*Ngule,*" the Kikamba word for "monkey," which is how the Hominid Gang described *Proconsul.* Wambua nodded gravely. That was what he had thought the tooth was, but he is a taciturn man, not given to boastful talk. I immediately began removing the matrix under the microscope the old-fashioned way, with a dental pick. Within minutes I could make out the nature of Wambua's find: It was the whole left half of the maxilla of the skeleton of the pothole *Proconsul,* still sitting there undamaged after 34 years. It fitted perfectly onto the skeleton in Nairobi when we got back, as I knew it would.

We carried on all week, collecting and inspecting lumps, washing them, sieving the residue, breaking the lumps down further, and cleaning matrix off of fossils when we could keep the airscribes going. That part of the work soon grew tedious. For a diversion, every few days we would prospect the surface of another site Louis and Mary had recorded on the island.

That first week we had several other important tasks to perform, too. Kamoya drove back to Mbita and called on the D.O., who was pleased that we had the courtesy and good sense to notify him of our presence and intentions. Kamoya invited the D.O. to come to our camp for dinner on Friday, to see what we were doing, and he happily agreed. One of Kamoya's other tasks was to buy a new dress for a lady whose *shamba* we had damaged in making a track on which to drive to the site. As worked out through the chief, this dress was her agreed-upon compensation for the maize we had destroyed, and she was pleased. He also brought back a goat, which we planned to serve the D.O. As I wrote in my camp journal:

> *Friday, May 18, 1984.* The D.O. came to supper, arriving about 7:30 in a most impressive way. He roared up the hill, nearly climbed the big rock in front of the mess tent, and came to rest between the table and Kamoya's tent. He didn't stop, but crashed along into the bushes where he eventually ground to a halt. He took out the guy ropes on one side of the tent. He said his brakes were bad and he had already hit a cow on the way up!
>
> As it was, he came with two Mr. Odiambos (one the big *askari* [or policeman]) and we all had a fine evening. The goat was superb. We talked about lots of things, looked at his brakes, lent him some petrol and they roared off on their way. Very pleasant people, who have offered to help us all they can.

Another priority was to get the camp running smoothly. Before we had been on Rusinga very long, we made the acquaintance of an enterprising lady named Rockette Ndege, who pronounced her Christian name with an emphasis on the first syllable "because I am so fast, like a rocket," she explained. Rockette thought a camp full of men offered her many opportunities for advancement. She offered to find us nice African wives for the duration of the dig, a proposal we declined since all of us had wives at home. Undaunted, her next offer was to keep the camp in water, chickens, fish, and goats, for a modest sum. Soon she and three other Luo ladies were each making five trips a day down to the lake and back up again. On the way up, each carried a full jerry can (20 liters) of water on her head. Paying Rockette and her ladies was much cheaper than taking the Land Rover down to the lake to get water, because petrol was so expensive. Using Rockette's supply service was also much easier for us in terms of labor and made for friendly relations with the locals, which we view as crucial to our success on any expedition. After all, we are always guests in an area and feel we should behave like guests, respectfully. It is an immense help to have not only the local officials but also a few ordinary people working with us because they can then explain to others, who might be suspicious of our intentions, exactly what we are doing in the neighborhood.

Early in June, Solomon came to me and told me there was another Dassenech on Rusinga in addition to himself and his brother Aila. This seemed improbable, for his homeland near Lake Turkana was a very long way from Rusinga, over 300 miles by road. When I mentioned the remark on the radio to Richard, he told me I must have misunderstood Solomon's Kiswahili. When he came up next, Richard questioned Solomon and got the same story: A Dassenech boy named Korobei was working on the *shamba* of a man called Opiato in Kamasengere, the next village. Richard knew that Opiato was a Game Department warden who had formerly been in charge of Sibiloi National Park, at Allia Bay at Lake Turkana. His suspicions aroused, he wondered if Opiato was keeping the boy as an unpaid worker. ("Unpaid worker" is a polite term for what a human rights activist would call a slave.) The next time Richard flew up, he went to question the boy and turned up an even odder story than he had anticipated.

Korobei came from Ileret on Lake Turkana. As was usual among the Dassenech, Korobei was sent out daily with his family's herd of sheep and goats from the time he was a young boy. His job was to guard the animals, look after them while they grazed, and take them to water when they needed to drink. The boy was a bit of a daydreamer and one day awoke from his reverie to realize that he could not see a single sheep or goat anywhere. He must have been frantic to find them, but after much searching he realized he had lost the whole herd: his family's total wealth. He had made an unforgivable mistake.

Korobei panicked, certain that he would be killed by his kin in retaliation, and fled to Opiato, whom he begged for asylum. Opiato agreed that Korobei was in danger of a serious beating or worse. Though Opiato was based at Allia Bay further south on the lake, which was about 75 miles from Ileret, he thought that was still too close to home for Korobei's safety. He could hide the boy temporarily, but gossip travels fast and Opiato feared that the family would soon learn where the youth was. The next time Opiato had leave—what an American would call vacation time—he smuggled Korobei out of Allia Bay and took him to his home on Rusinga.

Opiato gave Korobei the job of helping his wife and looking after his cows. Cows are much more costly than goats and Korobei had an appallingly bad record as a herdsman, so you might think Opiato was taking a terrible risk. Perhaps he thought Korobei had learned a lesson from his fright and flight and would now be more responsible. More likely, Opiato was confident that even Korobei couldn't lose a herd of cows on a small island. Another issue was that Opiato and his wife had no sons, only daughters. Looking after cows is not a girl's job in Luo society, not under any circumstances, and arranging for their care had probably always been a problem for Opiato. Now he had a young man to do the job and Korobei promised to work hard. In time, Korobei became like a son to Opiato and his wife and he married the best-looking girl on the island. He is expected to inherit Opiato's farm, since farms are not left to daughters. Korobei was a foolish boy from the north who lost a herd, but he found himself again on Rusinga.

I am squatting next to the relocated "pothole" site on Rusinga (R114), which must have looked much like this when Louis Leakey left in 1950. We cleaned off the vegetation and surface sediments to show its rounded shape. (© Mark Teaford.)

For us, Rusinga was teeming with opportunity too. We found more giant hyrax fossils, bits of *Deinotherium,* giraffids, rhinos, pigs, various rodents including springhares, anthracotheres (primitive ancestors of hippopotamuses), a fantastic turtle skeleton: all kinds of weird and wonderful Miocene animals. More pieces of the pothole *Proconsul* kept turning up as well as portions of a leg of the next larger species, *Proconsul nyanzae.*

We finally got the pothole itself entirely cleared off so we could begin digging into the sediments to see what, exactly, this mysterious structure was. Our plan was to cut a large wedge of rock out from the outside of the vertical pipe, to see what the surrounding sediments were like and to look for the bottom of the pipe, which ought to reveal whether the structure was initially a pothole or not. Once we could see the relationship between the contents of the pipe and the surrounding, or country, rock, we could determine whether they had been laid down at the same time or not. This was important since it was the gray country rock that had been dated to 18 million years, not the greenish rock that actually enclosed the *Proconsul* fossils; we needed to know if they were contemporary if we were to be certain of the antiquity of *Proconsul* and the other fossil animals. The work was difficult; the gray rock was very hard and indurated. The greenish pipe itself was tantalizing because we kept seeing more and more fossilized bits of bone as we exposed it inch by inch.

The air compressor was an enormous frustration. It shook itself to pieces regularly and the vibrations snapped part after part that had been badly made or improperly assembled, as documented in my field diary from 1984:

> *Tuesday 22nd May.* At about 11 o'clock the compressor starting surging and eventually began to quit. There was only one bolt left holding the engine to the frame. The design is pathetic. The bolts were welded to a [metal] plate, but didn't pass through it! We took it apart and went to Homa Bay. There we had to wait for the *duka* [shop] owners to have lunch and then we got the parts fixed properly (if crudely).
>
> *Thursday 24th May.* We four began airscribing, but about 10 o'clock the pressure suddenly fell by half. We tried to fix it but to no avail. We think one of the valves in the compressor pump is not working properly. It's blowing back through the air filter on the intake, which it shouldn't. . . . A call to Holman's about the compressor gave little help, saying that it was either a leak or the regulator! Baloney! We'll tell R.E.L. [Richard] to bring big spanners [when he flies up] and [we'll] take it apart. At least we found that there is no service agent [for Holman's] in Kisumu. That would have been a wasted trip.
>
> *Sunday 27th May.* R.E.L. took the head off the compressor pump, which we found had a broken regulator valve due to in-

correct adjustment during assembly. We might get it back again before Wednesday [on a bus that regularly went from Nairobi to the field station of ICIPE, the International Centre for Insect Physiology and Ecology, in Mbita].

Thursday 31st May. The cylinder head hasn't arrived at ICIPE. Two calls to R.E.L. found that it didn't go the normal route to the storekeeper, so Kamoya went [to ICIPE] again. Unfortunately the place had a visit from the Minister for Agriculture and so he couldn't see the person who had [the regulator valve], the chief security officer. He went back to Mbita for 3 [P.M.] and didn't return with the part until 5:30 PM. We replaced it, not without problems since they had assembled the equalizing pipes backwards, but the compressor did have enough pressure and in fact blew the 110 lbs./sq. in. safety valve two or three times. Tomorrow we'll adjust it properly to about 90 and try to get the correct noise from it.

Wednesday 13th June. This morning started with Maundu, Kamoya and me taking apart the compressor. The starter [is broken] again. The casting had been breaking and pieces jammed the ball-bearings and they, in turn broke the plate, which jammed the starter. This accomplished, we found the machine would start, but not pick up speed. The bracket on the safety cage had broken and the cage caught on the flywheel pulley. That off, it was fine again. This is the list of breakdowns on a brand-new machine, just for my memory when I talk to the manufacturers:

1. Engine mounts all broke–2 days [lost] and a Homa Bay trip [to get spare parts].
2. After five days; cylinder head valves broken ~ 10 days down.
3. Leg fell off.
4. Starter broke.

5. Handle off.
6. Both brackets broken for regulator and safety cage.
7. Starter ratchet broken.

The constant troubles with the air compressor were positively maddening and the torrential rainstorms left us either dripping wet or steaming. On May 31, the storm was so bad that I recorded seeing a water spout develop near Mfangano, the next island, which is only a few miles away.

Soon our worries over the compressor and weather began to seem trivial as another problem arose. It began when Maundu skipped dinner once or twice and stayed out late without telling anyone where he was. On any expedition, it is essential to know where every member of our team is. If someone falls sick or is injured—twists an ankle, say, or gets bitten by a poisonous snake—we need to know where to look in order to rescue him. This basic safety measure of always having at least one other person know where you are has proven its value time and time again over the years. Although Rusinga was not as hostile an environment as East Turkana, where you can die of dehydration or a lion attack, everyone on our crew knew that you never go off alone without telling anyone where you are going.

Maundu had never been irresponsible before but soon his staying out became a habit and "late" turned into all night. The gang were all positive Maundu had a lady friend and Kamoya and I became concerned. Even though second wives are still fairly common in parts of Kenya, extramarital liaisons are not taken lightly. Besides, Maundu's actions reflected on our expedition and on the National Museums of Kenya. We always made a great effort to be fair, honest, and helpful to the people who lived and worked in the areas where we were fossil-hunting. We wanted to have friendly relations, to employ as many local people as we could, to buy local meat and produce if possible, and to give something back to the community. As a result, we very rarely had instances of theft or angry interactions with the locals, and we were able to return to the same areas year after year.

No matter who Maundu's lady friend was, his behavior was likely to cause trouble. If she was unmarried, she and her father might well expect Maundu to marry her. If she was already married, her husband was likely to be angry with this outsider who was romancing his wife every night. Besides, we all depended upon Maundu as we did on each other; we were a team and he was no longer doing his part. When he showed up late in the morning or too hung over to be useful, someone else had to do his work. There was an issue of respect and courtesy, too. As Kamoya said firmly, "If he has to be away he should have the good grace to tell us."

Kamoya had a long talk with Maundu and when he didn't reform, I had one too. It was an agonizing situation. We all respected Maundu and he was our friend, but he was behaving like a fool. He was jeopardizing his good job, the long-term future of the project, and quite possibly his health, as well as insulting his friends and colleagues. Kamoya and I kept trying to talk sense into him but he wouldn't listen. At the end of the season, Maundu quit his job at the museum and went off to join his brother in some get-rich-quick scheme that, predictably, failed. His health began to deteriorate and, a year or so later, we heard he had died of tuberculosis, which is a cause of death often given in Kenya when someone dies of AIDS. Whether it was tuberculosis or AIDS, this was a tragic and senseless end to my friend's life. I never knew what had so radically changed his attitude; he would not confide in me or Kamoya. I only know that an intelligent, gifted, and hard-working young man, the type of individual a developing nation sorely needs, was lost.

Many days I worked alone cleaning blocks from the pothole, a task which Maundu and I had once performed side by side in companionable friendship. In one of the blocks, I discovered the missing bones of the *Proconsul* skeleton's right hand; in another, a missing canine tooth. By the time the season was finished, we had once again increased the completeness of the *Proconsul* skeleton known as KNM-RU 2036. It was, then and now, the most complete early Miocene hominoid known from Africa.

With all the rain, the pothole itself was becoming a shallow mud bath. The deluges made a messy slurry of R114 that simply couldn't be worked, so the crew dispersed to search for fossils exposed by the rain elsewhere. At nearby Kaswanga, Nzube immediately found a primate maxilla, or upper jaw, and some teeth. Nzube is a small man with twinkling eyes that were still sharp despite his age. I never knew quite how old Nzube was, but in those days the gray had started to creep into his hair and the others teased him about being a grandfather. Nzube had always prided himself on his special ability to find monkey fossils, and he was a more than a little put out when his "primate" maxilla turned into a baby hyrax when it was cleaned.

On June 16, Nzube returned to Kaswanga with Wambua, to find real *ngules*—and they did. Nzube returned at lunchtime with a mixed set of primate teeth, and he and Wambua said there was a lot of bone on the surface. Martin must have seen the bones when he mapped the Kaswanga area previously, but he had marked the place as having fossil leaves and crocodile bones. Wambua and Nzube insisted the bones were primate. After lunch, we all walked down. Nzube led the way triumphantly and, typically for him, took us through every twist and turn, repeating every single step he had taken before finding the fossil teeth rather than going straight to the flat spot on the soil-conservation area next to millet fields. I recorded in my diary:

> We started looking and everybody started finding primates. There were hundreds of pieces—all primate. Carpals, tarsals, shaft fragments, teeth. It was amazing—we'd none of us seen anything like it. We came back to camp, muddy, because of the rain, but with a huge bag of bones. We washed them and found carpals, tarsals, and things no one had ever seen before. Even centrales! [Centrales are tiny hand bones that are very rarely found.] We thought [there were] 2 individuals, maybe three because of some milk teeth. The chief came to supper and we showed him what we were doing. He is very nice and helpful,

Peter Nzube (left) and Kamoya Kimeu (right) at the moment that we realized the extraordinary number of primate fossils that were preserved at the Kaswanga Primate Site on Rusinga. (© Mark Teaford.)

> but we were all wanting to do the bones, so as soon as he'd gone we got the pressure lamp out on the table. By the time Kamoya returned [from taking the chief home] we had sorted a fair bit and glued some pieces. We were all shocked, but especially Mark, since he'd no experience except what he knew of the literature. We went to bed the latest we had [since arriving], hoping for no rain.

The next day, incredibly, the finds continued. We began to take wheelbarrows of the loose surface sediments down to the lake to wash them through sieves, not far from a group of four hippos who looked on curiously and seemed to be laughing at us. We were laughing too because we were having such success. Sieving is normally a necessary but uninspiring job; this time, the sievers reaped bags of fossils.

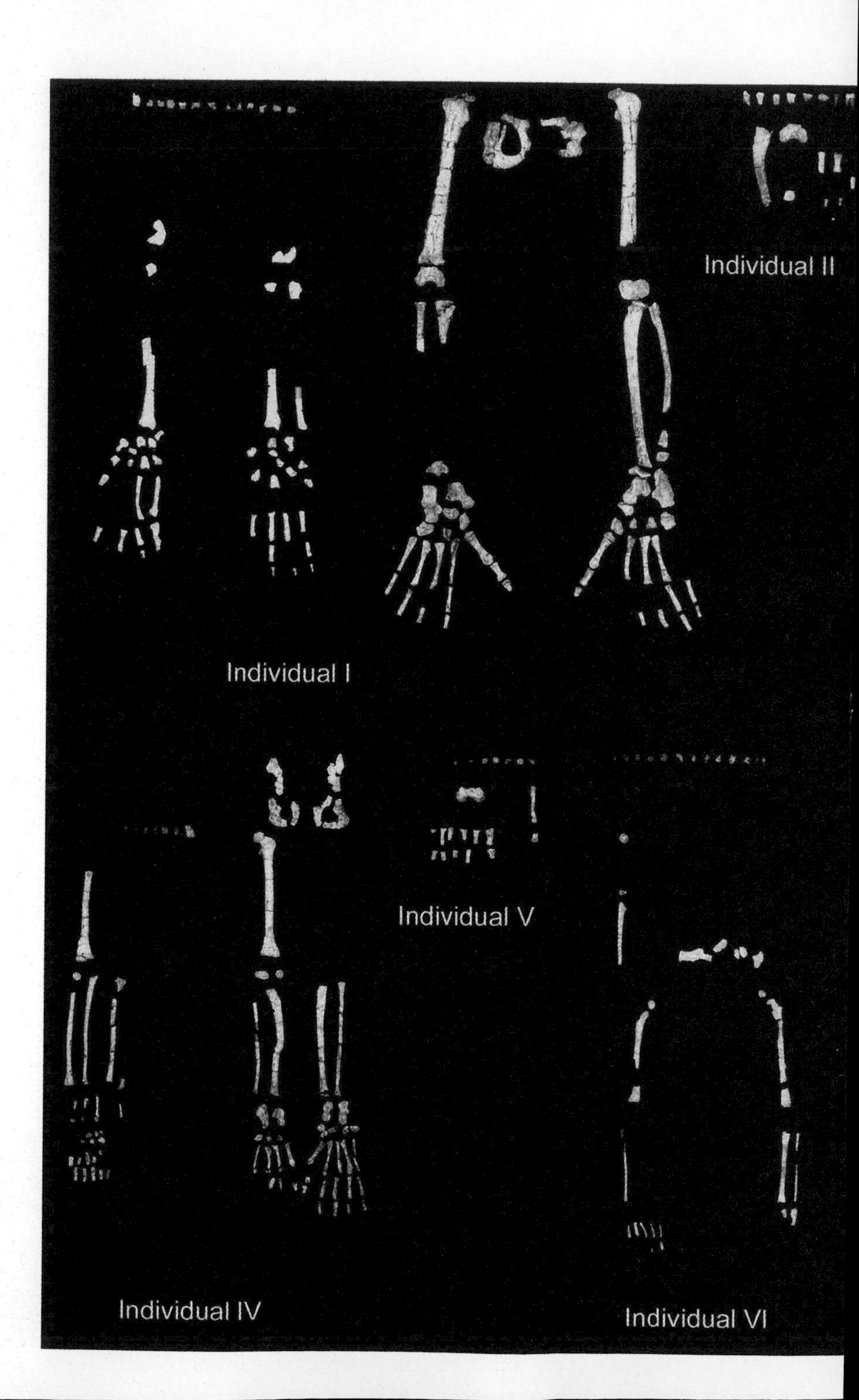
Individual II
Individual I
Individual V
Individual IV
Individual VI

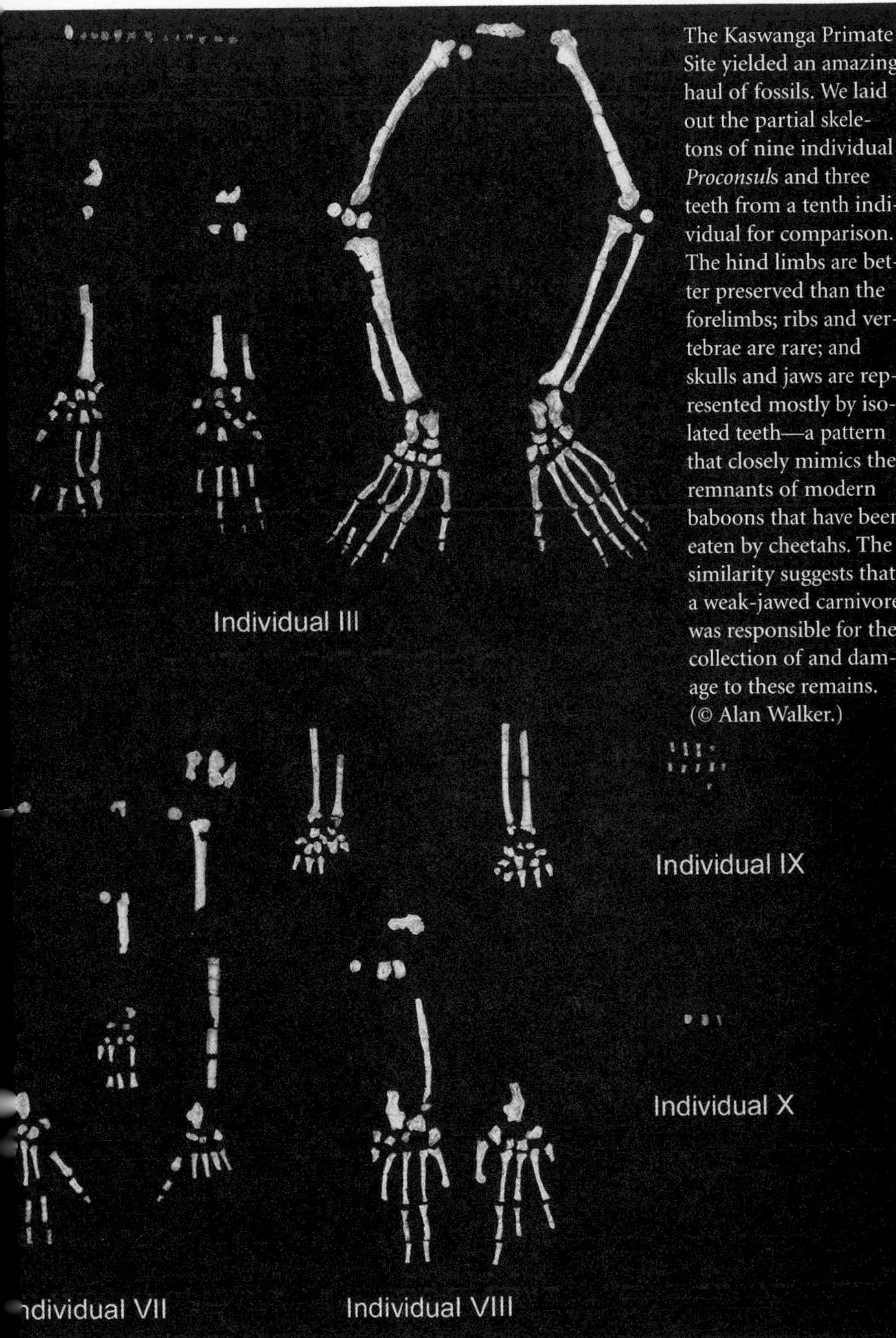

The Kaswanga Primate Site yielded an amazing haul of fossils. We laid out the partial skeletons of nine individual *Proconsuls* and three teeth from a tenth individual for comparison. The hind limbs are better preserved than the forelimbs; ribs and vertebrae are rare; and skulls and jaws are represented mostly by isolated teeth—a pattern that closely mimics the remnants of modern baboons that have been eaten by cheetahs. The similarity suggests that a weak-jawed carnivore was responsible for the collection of and damage to these remains. (© Alan Walker.)

Then Nzube found bones that hadn't washed out yet. After excavating a bit, we found they were the lower leg bones, both tibia and fibula, of an infant *Proconsul.* I started excavating them carefully and soon found its complete foot was next to the leg. Working with these fossils was difficult because the clay-rich soil in which they were embedded was cracking into tiny pieces as it dried. The bones were wet too and seemed stubbornly resistant to taking up Bedacryl, the liquid preservative we use, but if I didn't go slowly and soak each bone with Bedacryl before moving it, the fossils would break into minuscule bits just like the soil. Before I'd even finished the baby leg and foot bones, we found more: a squashed but larger tibia and fibula, and at its end, an adult right foot—complete. Two complete feet, attached to lower legs? That's the sort of find you never make—almost never anyway. But minutes later I found another foot, the left, and it too was complete and so close to the first adult foot that they were almost touching.

I decided to remove these two adult legs and feet in a single plaster jacket to preserve their spatial association and have time to take great care over them. I removed the few bones that were loose on the surface, after taking pictures, and then dug around the two legs to leave them supported by a thick pedestal of clay soil. Then I applied a layer of damp toilet paper to the top of the fossil and rock as a separator, to make sure the subsequent layer I was going to apply (plaster-soaked gauze) would not stick to the fossil. Finally, I swathed the unsightly lump with enough plaster and gauze to make a walking cast on a human leg. Once the plaster jacket had set dry and hard, I severed the pedestal and lifted the whole thing up to take back to camp. The plaster would hold everything in place and protect the bones until I had time to excavate carefully and slowly.

For days, the bones kept coming like rabbits plucked from a top hat by a magician. Amazingly, ridiculously, we found part of an adult hand, its bones in anatomical position, lying between the adult legs and the baby leg and foot. Though the hand bones were filled with tiny rootlets, I was able to rescue the fossils intact without losing the relative positions of the bones. Not only the major bones but even the tiniest ones were there, the little rounded sesamoid bones that ease the passage of tendons across joints. Although some bones were crushed and flattened, they had beautiful anatomical details that would reveal a great deal about these ancient apes and how they moved through the trees.

We were beginning to think about sudden catastrophes that might kill and bury several primate individuals of different ages in this way. Had we stumbled upon a primate Pompeii? We sent for our lunch at midday because we didn't dare leave these fragile, exposed fossils for a moment. As we freed them from the soil, we took all our treasures to one place where we could sit and slowly perform our rescue operation.

Nobody had ever seen anything like this before. Looking up from excavating and glancing around, I suddenly realized how much more bone there was on the surface. The area looked like the aftermath of a tickertape parade; there were little white flakes everywhere, and every flake was a small bone or a piece of bone. More amazing still, every bone we had identified so far was primate. We had stumbled upon a once-in-a-lifetime site.

There were too many fossils to take out in this season with proper care; we could already see that and we hadn't finished with the pothole yet. At that juncture, Chief Nditi happened to come by. We took a rest and explained to him what we were finding. He could see it was important to protect these remarkable fossils and thought it would be appropriate for us to put up a fence to protect the site until the next season. Then he went off to talk to the people who lived nearby, to get their cooperation and agreement. What we asked the chief to do, if possible, was to keep the site area out of cultivation for 5 years so that we could excavate. After that period, the site could become a maize field. The chief persuaded the locals that the plan was a good idea and Kamoya went to buy fencing material the next day.

We didn't return back to camp until five o'clock, and by then our heads were spinning with the extraordinary finds we'd made. There was every chance that much more was lying there in situ, waiting for discovery. We already had some pieces of sacrum and what looked like tail vertebrae, but we hadn't yet found any vertebrae from the trunk. We had found lots of teeth, too, but not many pieces of skull and mandible . . . yet.

That afternoon, we took as much material as possible back to camp and quickly washed it off. It was a fantastic sight, all those specimens spread out on the work table: bones, bones, and more bones, many body parts never seen before for any *Proconsul.* Then we went down to the lake to swim, because we were hot and dirty and our backs were aching. Mark had been sitting over the sieve all day, Kamoya and I had been excavating, lying or sitting down awkwardly, and the other men had worked hard too, carrying sieving material and prospecting for bones. I had the added discomfort of sunburned legs; I have fair skin that freckles and burns badly and I was so excited about the fossils that I kept forgetting to keep my legs covered or to add more sunblock. I didn't care; sunburn was a small price to pay for such fossil treasures.

Before supper I gave the gang a bonus, as I always do when they turn up something special, and they all went off to celebrate. After dinner, we got the lamp out to keep working with the bones in the dark. Trying to light as large an area as possible, Kamoya put the lantern up on the lid of the milk churn we use for water. Just as we decided we were too tired to concentrate any more, the lamp slid off the milk churn, breaking the glass, which cost 80 shillings (about $10 in those days) to replace. We put the bones away and cleared up the broken glass and went to bed very early. I scribbled the day's events in my journal and fell sound asleep, only to wake up when the men returned at about ten o'clock, very merry and still excited. They crashed into Kamoya's tent and kept him up late with stories and tales. Nzube decided he would have a cable address that said simply "*Ngule*—Nairobi." What a day!

The next day Mark and I got started identifying, cleaning, and gluing the bones while Musa and Aila swept the area and sieved the loose sediment. By the end of June 18 we knew we had at least four individuals: the baby, an older infant or "child," a young adult, and an older adult. What we had was an excellent set of partial skeletons from one place and time showing how growth had proceeded in *Proconsul,* a real rarity among fossils. For the first time, we could measure complete lengths on many of the bones and would be able to look at how *Proconsul* grew from infancy to adulthood. We had discovered another site on Rusinga that would be famous among paleontologists: the Kaswanga Primate Site.

Before he managed to get the fencing, Kamoya had a visit from a woman who deeply objected to our taking over the primate site. It was difficult to understand exactly what her concern was, but she was very upset and Kamoya went with her to the chief. Chief Nditi listened for a while and then told her that he was angry that she'd stopped the work, because her interference showed that she didn't believe the village elder or Kamoya when they promised that proper compensation would be paid. He told her sternly to be quiet or go to jail; he'd put a fence around all the land if she didn't shut up. She was suitably subdued by that threat.

After we gleaned everything from the Kaswanga Primate Site that we could, we returned to the "pothole" that had been our initial objective. As the ground dried out, we continued to dig out the wedge-shaped section next to the pothole. The pothole hypothesis faded rapidly as we began to be able to see the shape of the structure clearly. Though the main pipe was still going deeper, its diameter seemed to be increasing. The gray country rock was arranged in matching layers all around the pipe. There were even some slickensides—polished marks in the rock created by the movement of one layer against another—that showed how the sediments had compressed around the pipe. What we had was the remains of a buried tree trunk!

Our scenario was this. While the tree was alive, a sediment-laden river or stream had deposited the gray sediments and buried the tree

When we excavated down through the rock surrounding the so-called pothole at R114, we could see it was an infilled tree trunk, not a pothole. (© Alan Walker.)

trunk to a depth of at least 12 feet (4 meters). We could tell that the tree—or some cylindrical object—had been buried because the sedimentary layers on all sides of the object were the same. We could even deduce the direction of the stream because pebbles and larger rocks collected on the upstream side of the tree, while the finer sediments filled in around the trunk and downstream of it. The tree died; its trunk rotted, leaving a useful hollow that was a cozy den for monitor lizards, pythons, bats, and small carnivores; when they died, they left their skeletons inside the tree hollow. Some of the carnivores probably brought prey in too, perhaps to feed to their young. Some of the *Proconsul* bones had clear toothmarks on them from a small carnivore of some sort. Over time, the hollow filled up with bones and sediments that solidified into the greenish bone-filled rock that Whitworth first found.

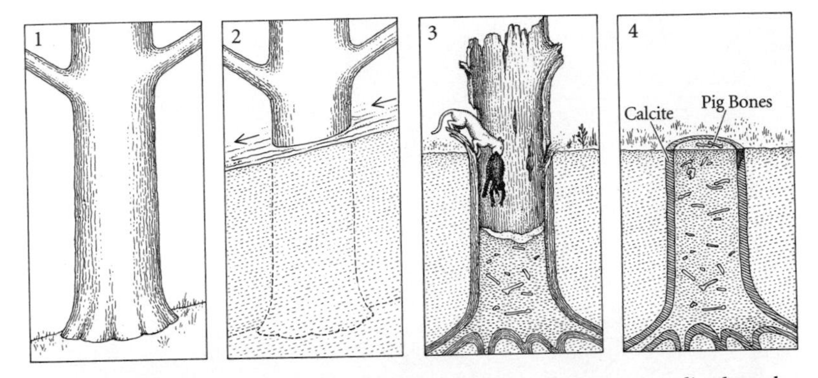

In our scenario, the R114 site formed like this: (1) A rain forest tree lived on the flanks of Kisingiri volcano about 18 million years ago. (2) An eruption of volcanic ash and debris buried the tree up to at least 13 feet (4 meters). (3) The tree died and its trunk rotted away, leaving an inviting hollow. Primitive carnivores denned in the hollow and brought in small animals that they preyed upon, including *Proconsul.* (4) Erosion uncovered the infilling of the tree trunk and exposed the fossilized skeletons in the sediments. (Reproduced from Alan Walker and Mark Teaford, "The Hunt for *Proconsul,*" *Scientific American,* 260, no. 1 [January 1989]: 82. Diagram by Tom Prentiss 1936–1997.)

Once we understood we had a fossilized tree trunk, we began to realize there were some other, roughly circular deposits of green rock locally. These too were probably buried trees, meaning that R114 was part of a forested area. We might easily find fossilized bones in the other green deposits, too.

As well as refuting the pothole hypothesis, learning that the site was a hollow tree proved that the bones and the rocks contained in the tree were contemporaneous with the gray country rock. We knew from radiometric dating that the gray country rock had formed 18 million years ago, so the tree was alive and was buried by these sediments at about that time. Within a few years of burial, the tree trunk had rotted and began to fill with the skeletons of various animals. Radiometric dating of rocks this age is not precise enough to detect the passing of a single year or two; for Miocene rocks it is rarely accurate to less than 100,000 years. In geologic terms, then, the animals lived at the same time that the sediments first buried the tree.

Once we realized that the so-called pothole was actually the infilling of a former tree trunk, we began to see others like the one I am standing next to in this photo. The area around site R114 must once have been forested. (© Mark Teaford.)

Because we understood how the primates and other fossil animals came to be in the hollow tree—some lived there and some, including the primates, were carried in by small carnivores—we could also be fairly sure that *Proconsul* had lived near the site. Small carnivores tend to have fairly small territories and besides, they simply couldn't have carried whole, adult *Proconsuls* very far. This proved that the fossils represented a local animal community that lived near the tree at a specific time during the Miocene. From a beautiful fossil assemblage like this, so well preserved and so firmly dated, we could build a detailed and accurate picture of the ecology of the area as well as of the species *Proconsul* itself. It was beyond all reasonable expectations and we were glowing with excitement.

We had found out a great deal about Miocene apes in one short season—and there was more to come.

6

An Embarrassment of Riches

We weren't finished with the Miocene sites yet. We were back at Rusinga in March and April of 1985 and returned again in May and June of 1986 and 1987.

The basic crew from the museum stayed fairly stable over those years, with only a few changes. Martin left the museum and the expedition, as did Maundu. Benson Kyongo, a young Wakamba man, first hired as a cook's assistant in 1984, now joined the Hominid Gang and soon showed the intelligence and personal qualities that made me think he might be Kamoya's successor. Benson had finished high school with good enough grades to go to the university, but he didn't get a scholarship and his family could not afford the fees. Luckily for us, he started working at the museum and excelled, taking opportunities to further his technical education and skills whenever they were offered. Mark stayed on at Hopkins as an assistant professor and came out to dig every year; Blaire Van Valkenburgh, my next postdoctoral fellow at Hopkins, came for the 1985 season; and Chris Beard, Audrone Biknevicius, and Hannah Grausz—then graduate students at Hopkins—each spent time working with us. In some of the final years our grant paid for an internship at the National Museums of Kenya for a young Luo man named Iziah Odhiambo Nengo. He participated in the fieldwork as part of his training.

Word had gotten around about our extraordinary site, so we began to have visitors. When she could spare the time and get transport, my wife, Pat, visited from Nairobi, where she was doing her own research. Louis Leakey had died in 1972 and never knew of the fossils my team found, but Mary was still alive and came to see our site. To my surprise, Mary barely recognized "her" tree, under which she had found the little *Proconsul* skull when she saw it again. I have rarely known Mary to forget much having to do with excavating and fossils she was involved with. To be fair, she was 72 years old at the time and had a patch over one eye due to a medical problem, so her vision was not as acute as usual. Mary being Mary, a little thing like an eyepatch was not going to keep her out of the field when there was a site and good fossils that she wanted to see.

Other paleoanthropological colleagues from around the world who had reason to be in Kenya came for brief visits too, because by then the Kaswanga Primate Site and the *Proconsul*-in-a-tree were well known. Richard came up from time to time but always briefly. On one visit, he negotiated with Chief Nditi to build a site museum on Rusinga if the chief could find three acres to build it on, so that the people could learn about their past and the fossils that had been found on their island. It seemed only right that the people who were in this historic place should have a way to learn about the discoveries made there.

We regularly camped under the lovely fig trees near the Kaswanga Primate Site, where we could hear the fish eagles flying high overhead. Dozens of the beautiful black and yellow weaverbirds chattered among the leaves and hung their cleverly constructed nests from the figs' branches to keep their young safe from predatory snakes, some of which are remarkably good at climbing trees. Tremendous nightly storms continued to occur, but we were smarter about placing our tents, digging drainage trenches, and checking for leaks during daylight hours instead of waiting until a catastrophe happened in the middle of the night.

During the second field season on Rusinga, we made the acquain-

At the Kaswanga Primate Site, Blaire Van Valkenburgh battles sunburn by wearing socks on her hands. (© Mark Teaford.)

tance of a strange species of cricket, about 4 inches long, that lived in holes in the ground all around our camp. In the evening these crickets chirped continuously, making a terrible din commensurate with their exaggerated size. Blaire was the first among us to crack. "I know," she said with a rueful grin, as she put stones over the openings to the crickets' holes, "and I'm an animal lover, too!" But none of us could sleep through their racket. I found, to my amusement, that Felix Oswald had met with similar crickets on his brief trip to the region in 1911 and wrote of a cricket that "fills the air with an extremely shrill and penetrative keening that is most fatiguing to the ear."

Blaire had the worst time of all of us with the strong sunshine, as she is a very fair-skinned redhead. She wore a hat, long-sleeved shirts, and pants, but still burned her wrists and hands. Finally she took to wearing socks on her hands while she excavated. She looked ridiculous—she was the first to laugh at herself—but her unorthodox costume was effective. Besides, she loves fossils and excavating and never complained.

That second season the amazing flow of fossils continued undiminished. Our first full day in camp after setting up was a Sunday, normally a rest day, but I got bored after watching the birds awhile. "I'm going to stroll over to the fossil site," I said to Blaire and Mark. "Do you want to come?" They did. We were ambling down the path and suddenly I saw something white on the ground. "Primate," I said. The others looked around, astonished that I could spot a rare fossil as I was simply walking along. When I stooped down to pick up half a *Proconsul* finger bone, I suddenly saw four more hand bones. Then all three of us were on our hands and knees, picking fossil primates off the surface as if we were picking ripe berries from a bush. We went back to tell Kamoya what we had found and to get a sieve. By the end of that afternoon, we had found part of the first hand and wrist of *Proconsul nyanzae* ever seen. If that had proved to be the only good find of our whole field season, we'd have counted it a success. But on the third day some of the men started working on the Kaswanga Primate Site and immediately started finding baby and juvenile phalanges of *Proconsul* to add to the collection we had made the first year.

From then on, my field diary shows a daily log of new *Proconsul* bones and teeth from both *P. africanus* and *P. nyanzae,* not to mention other types of animals. There were carnivores, which may well have been responsible for the great accumulation of bones in the fossil tree; the peculiar chalicotheres, with their elongated, horsy heads and big clawed hands and feet; insects and leaves; antelopes; a tiny bushbaby fossil (bushbabies are primates too but they are more closely related to lemurs than to apes or monkeys); a rare fossil bat skull; a chameleon skull; and many rhinoceros fossils, to name a few.

Near Mary's tree, where the *Proconsul* skull had been found in 1948, Wambua found a maxilla that he was certain was *Proconsul.* Kamoya insisted it was a hyrax, though, one of those peculiar giant ones we had found on Rusinga the year before. He was most embarrassed when I pronounced the maxilla to be a *Proconsul,* and I almost laughed at the "I *told* you so" look Wambua shot at Kamoya. Kamoya then suffered a disheartening period of not finding any good fossils for a few days, probably because he was feeling glum and lost in the Miocene, where he didn't know every species as well as he did in the more recent Plio-Pleistocene. That same day, some of the gang found some pieces of jaw and teeth from an enigmatic carnivore called *Teratodon* that, judging from its odd, rounded, buttonlike premolars, specialized in eating and crushing snails. There is nothing on earth like it alive today. The best I can do to describe it is to imagine something like a sea otter crossed with a crab-eating mongoose and a hyena. The Miocene was a very peculiar time full of strange and sometimes unrecognizable animals.

Only a few days after we got settled at Rusinga in 1985, our old friend Rockette Ndege turned up, offering to keep us in water, chickens, fish, and goats once again. She was as friendly and energetic as ever and it was good to see her. She was so vivacious that Kamoya and I exchanged glances, suddenly wondering if she had been Maundu's girlfriend the year before. We didn't ask and Rockette never mentioned him. Every time we went back to Rusinga to excavate over the next few years, Rockette arrived almost immediately to offer her services once again and we were happy to employ her and her women.

During the 1985 season, the renowned geologist Richard Hay came to double-check the work on the Rusinga stratigraphy and geology for us. He had sorted out the geology of Olduvai Gorge for Mary many years before and they were firm friends, so Mary was pleased to find Dick in our camp when she came to visit for a few days. Dick concluded that the mapping and interpretation done earlier were basically correct, needing only a few revisions. He and Mary were able to relocate a site where she and Louis had found fossilized caterpillars years before, too.

Mary and Dick were in camp when an unexpected event woke us all up early one morning. There was a huge commotion down at the shore below us: yelling and shouting and all sorts of ruckus that surpassed the normal Luo exuberance. We spotted some men in a canoe

Mary Leakey listens while Dick Hay and I discuss the geology of Rusinga in 1985. Mark Teaford is standing on the crest of the hill. (© Blaire Van Valkenburgh.)

who were the source of most of the noise; they seemed to be having a problem with hippos and were trying to scare them away. The danger posed by hippos to people in small boats should not be underrated, as Pigott had learned to his cost many years earlier. From our position on the hill above the lake, I could see one of the hippos galloping on the bottom in water about 4 or 5 feet deep, looking for all the world as if it were performing the hippo butterfly stroke. Finally we realized that the real problem was a dead hippo that the men wanted to retrieve.

A hippo, which weighs in at about 5,000 pounds (or 2,272 kilograms), is an enormous amount of meat for people who are often short of protein. Even though it was dead, this hippo was being guarded ferociously by the other hippos from its herd. They didn't want to leave its carcass. By the time we got down to the shore, the men had managed to drive the live hippos off far enough so that they could drag the carcass into a foot or so of water near shore. A crowd was gathering but, oddly, it consisted of perhaps thirty men, an assortment of old women, and some members of our expedition. Why were there no younger women there, like the energetic Rockette, for example, who would surely want a share of the meat to feed herself and her children? We noticed that, as the crowd grew, the men kept looking pointedly at Blaire. She seemed to be drawing more than the usual attention a red-headed, white woman attracted on Rusinga, but we couldn't figure out why. Finally, the shore became so crowded it seemed as if the entire population of the nearest village, Kamasengere, was milling around except that there were no young women. Then Chief Nditi arrived and gave his permission for the butchery to begin. Instantly, bloody pandemonium broke out. I described it in my diary:

> People are running everywhere with knives, big bloody hunks of meat, strips of skin and fat. Some are waving *pangas* [machetes], others on boats are sailing off with big bits, like legs etc. Later we learned that the game dept had shot the hippo some time ago, but that it had taken a long time to die. It was a very old male. [We learned that] women are not supposed to see the hippo's tail or they'd have either children with tails or none at all. Hence the old women and Blaire [who knew nothing of this prohibition] were the only ones to come near! The chief had a canoe to take a thigh and the huge, heavy head to his home.

It was a miracle that nobody chopped the leg or arm off a fellow human in the frenzy. Later I saw an engraving of a similar episode witnessed by Sam and Florence Baker, the famous Nile explorers of the 1860s, and I realized that hippo slaughtering had remained unchanged by the passage of over 120 years. Then, as now, meat was meat and very valuable to undernourished people.

The next night, we had the storm to end all storms. I woke up when my tent, with me and my possessions inside, began to lift off the ground. I was afraid that the tent and I were about to be blown

In 1861, Sam and Florence Baker witnessed the enthusiastic butchery of a hippo. When we camped on the shores of Lake Victoria 124 years later, the scene was much the same. (From Samuel Baker, *The Nile Tributaries of Abyssinia* [London: MacMillan, 1886.]; © Pat Shipman.)

off the cliff into the lake. Suddenly I remembered that the aerial photos of the site were in my tent. The location of each individual fossil is marked on the appropriate aerial photo with a pinprick and its number; the aerial photos are the best and most important documentation of the place each fossil is found. If that information were lost in the storm, all the provenience information on every single fossil we had collected since we started work at Rusinga would be lost too. As I tried desperately to hold the tent down from the inside, I gathered up the photos with one hand and stuffed them in the safest place I could reach: under the mattress. I yelled for help, too, and soon others rushed over to save me and the aerial photographs. Musa and Benson were trapped inside their own tents, which had collapsed in soggy messes on top of them as they slept. Blaire was trying to save some of her things, since the roaring wind had ripped a huge hole in the roof of her tent, soaking everything. Some of the men were trying to take down the fly sheet that kept the food store dry before it was shredded to pieces, and still others were trying to find someplace dry to shelter the food supplies. The booming thunder shook the very bones in our bodies and almost continuous flashes of lightning lent a hellish aspect to the scene.

In the morning, all was quiet and the world seemed exhausted. The beach from which we had bathed every day was unrecognizably reshaped. Hundreds of dead fish lay washed up everywhere along the shore; branches, leaves, reeds, and other vegetation lay scattered about like discarded rubbish. The afternoon before, we'd covered the excavation with tarpaulins held up by poles and had laid plastic directly on the ground and weighted it down with stones, but our precautions were useless. We could see a good foot of mud and water on top of everything. We removed the drenched and heavy coverings, bailed the excavation out as best we could, and dug trenches to lead the water away while the ground steamed itself dry under the hot sun. An exquisite fossil leaf-bed that Mary and Blaire had been excavating meticulously and slowly was destroyed; no paleobotanist would ever study those leaves. A pair of fragile rabbit skulls we had been working on were gone forever, with no scrap left behind. The only good news was that Rockette brought us some lovely fresh fish for dinner.

One of the strangest incidents during our time on Rusinga happened on a rest day, while Blaire and I were taking a long walk to watch birds. Rusinga has a marvelous variety of birds that includes the strange hammerkop, an awkward-looking, long-legged, brown bird with a squared-off crest on its head. The hammerkop builds gigantic, ungainly nests that resemble a random pile of sticks, often with plastic bags or old clothes tacked on. The result doesn't look from the outside as if it would offer any protection to eggs or chicks, but apparently it does. We were hoping that day to come across some hammerkops. What we found instead was the island's resident madman, who leapt suddenly out of the bushes at Blaire, gesticulating and mouthing incomprehensible but heartfelt words. He was very old and very dirty, with a long, unkempt beard. He wore only a few skins tied around his shoulders, instead of the Western clothes that everyone else on Rusinga wore. His hair was matted and his eyes were wild, but what really shocked me was that *he was in chains.* They wrapped around his waist and connected to manacles on his ankles.

I had never seen anything like this in all my years in Africa. Later I asked the local people who he was and why he was chained up. They explained that he was very old, so old he had been a soldier in World War I. When he returned from the war, whatever he had seen and experienced drove him mad and he persisted in molesting and raping girls and women. I wondered if he been in one of the terrible, nightmarish battles: the Somme, perhaps, or Passchendaele. Had he suffered from shell-shock, like the World War I poets Wilfred Owen and Siegfried Sassoon? I couldn't tell, not speaking his language, and I knew that psychiatric help would have been completely unavailable to him on Rusinga. His family chastised him and tried to persuade him to behave himself properly, but the poor man was beyond reason. In response to pressures from the community, and to fulfill their responsibilities as best they could, the family put chains on the former soldier so he couldn't run after the girls.

For decades, the pitiable ex-soldier was mostly confined to a dark hut, where someone took him food daily. As he aged he improved, or perhaps he simply became less agile and so less dangerous.

When we met him, he was about 90 years old, so he was sometimes let out to look after the family's cows. It was a hard lesson in the realities of life. Even in Europe, which prided itself on being civilized, the shell-shocked soldiers were inadequately treated and were often deemed a shameful embarrassment to their families. How much worse had it been for this unfortunate man?

Ever since our first field season on Rusinga in 1984, we had wanted to follow up on a discovery on Mfangano, the next-door island on Lake Victoria with a name that is unpronounceable by most Westerners.

Both Rusinga and Mfangano Islands are remnants of the same ancient and alkaline volcano, Kisingiri. Like the fossils on Rusinga, those on Mfangano are preserved courtesy of the strange chemistry of this long-dormant volcano. On both islands we could recognize the same geological layer, the Hiwegi Formation, as the source of most of the fossils. We were sure that the fossils from Rusinga and Mfangano were contemporaneous with each other and therefore so were the two species of *Proconsul.*

The reason Musa had been on Mfangano in January of 1984 was that Martin had been gazetting the island's sites and had taken Musa along as his assistant. On that occasion, Musa had found a nearly complete right femur of a *Proconsul nyanzae,* as well as two large bones from its right foot. He thought there were probably more bones, too, so during the 1984 season we had borrowed a small dhow from one of the teachers on Rusinga and sailed across for a quick reconnaissance. I wrote in my diary:

> *Thursday 14th June, 1984.* Four of us (Musa, Kamoya, Mark and I) went to Mfangano today. We left at 8 A.M. in the teacher's boat from Tambasa and with good winds, were at our destination in about an hour and 15 minutes. We didn't quite land where we were supposed to . . . because the captain and mate of the canoe didn't know much Swahili and a Luo lady traveling with us confused them. However, it wasn't far from where we wanted to go—about ten minutes walk. Kamoya, who is afraid of a lot of water, was pleased to be on land again. Musa found the spot quickly, despite it having been the dry season when he went in January. The [site] is a most unpromising place, a piece of bush in the middle of small *shambas.* It is so flat and covered with old stones, you wouldn't suspect anyone could find anything there. Anyway, we cleared the big rocks and started clearing the soil, hoping for *in situ* bones, such as the tibia between the femur and ankle. However, what we found were two vertebrae: one last thoracic and another lumbar, not adjacent [to each other in the spine]. They were very well preserved. . . . They had both been chewed in such a

way that the [vertebral] column had been disarticulated. . . . The sieve had a bit of sacrum in it—one of the first from any African Miocene hominoid. We decided that it would need at least a week [to excavate the area properly].

We intended to go back to Mfangano in 1985 for a longer look and excavation, but we couldn't. There were just too many fossils on Rusinga to be recovered that season.

When we started the field season of 1986, doing justice to Musa's *P. nyanzae* site on Mfangano was the first item on the agenda. This excursion necessitated some of the most complicated logistics for getting a crew into the field that I have ever been associated with. Being an island, Mfangano is an inaccessible place to get a good-sized crew, camping gear, equipment, and supplies to. Although Rusinga, where we had worked for 2 years at that point, was also an island, it was connected with the mainland of Kenya by a long causeway that had been built out across the lake. We could drive trucks and Land Rovers to Rusinga with relative ease. In contrast, there were only three ways to get to Mfangano: by flying, which was not a practical way to transport a lot of heavy material and many people; by the ferry, a small steamer named the *Kamongo* (meaning lungfish), which could carry heavy goods from Mbita Point on the mainland to Sena on the island, but which was in demand by the local populace; and by small boat. We used all three methods in a carefully choreographed scheme.

We started with two teams. The advance party consisted of Mark and most of the gang, who drove from Nairobi in a truck and a Land Rover loaded with gear, including a 20-foot boat. They drove to Mbita on the mainland shore of Lake Victoria, where they unloaded our huge piles of camping gear, equipment, and supplies. After unloading, the truck (still carrying the boat) drove across the causeway to Rusinga and along the single road that encircles the island. The boat was left at the house of a schoolteacher, Ronald Ouma, who lived at the end of Rusinga nearest Mfangano. Ronald didn't want to look after the Land Rover while we were on Mfangano for some reason; perhaps he thought it was too valuable and would be stolen, and he would be blamed. The truck returned to Nairobi while the Land Rover stayed at Mbita at the station of ICIPE, with which we had excellent relations. Once the boats and vehicles were settled, the people, gear, and supplies had to be loaded onto the *Kamongo* to go to the village of Sena on Mfangano Island. The second, aerial, team consisted of Kamoya and me. On the fifteenth of May, Richard flew us to Sena to meet the others.

Bush airstrips are primitive affairs by Western standards, usually being somewhat flat pieces of ground marked off—if marked at all—by rocks covered with whitewash. They have no tarmac, no lights, no buildings, none of the facilities that normal airstrips have. The Mfangano airstrip, however, was really awful even by the standards of African bush strips. There was a gate across one end—get that plane up fast, Richard!—and deep erosion gulleys criss-crossing the strip. In the very middle of the strip was a cement-hard termite hill several feet high, guaranteed to remove the landing gear from any plane that hit it. Alongside the strip, in case the pilot had not aligned the plane perfectly for takeoff or landing, there were telephone wires to entangle the plane. The whole airstrip was on a slope, too.

Richard is an exceedingly careful bush pilot and took three low passes over the airstrip to check it out before he attempted a landing. He admitted later that he wasn't very happy with the condition of the strip and his blood pressure went up a lot during landing. Afterward, he was eager to leave, but more because he just doesn't like Rusinga and Mfangano than because he was worried about the runway. Time wasn't going to change the runway much. The team was waiting for us at the strip and helped off-load all the perishable and heavy supplies that we had brought on the plane: enough to keep us going for a week or so. Then Richard flew Benson and me back to Rusinga and another bush strip in better condition, so we could go to the schoolteacher's house to collect the museum's boat.

The boat proved to be incredibly heavy, though it was only a 20-

foot whaler with an outboard motor. We had to enlist a half a dozen local men to help us get it to the shore down a steep path that was helpfully bordered with cactus plants. Richard showed me the tricks of the engine and then left to fly back to Nairobi. Benson and I set off in the boat to chug back to Sena: an open-lake crossing of more than 4 miles. At Sena, we ate lunch, packed up a few of the people and some of the gear, and set off around the island's shore toward the area where we would camp, near Ramba School. For 5 long hours, I was a taxi-boat driver: loading up the boat at Sena, driving it to Ramba to drop off the people and gear, and then turning around to return to Sena to get another load while the people already at Ramba started setting up the camp. Ferrying everyone and everything around took six slow round trips. For some of the trips the boat was so heavily loaded that its gunwales were just above the lake's surface. I worried the entire day because I knew none of the men could swim well and some of them couldn't swim at all. If anything happened, I'd be the one to try to save them and myself. I also kept a sharp eye on the weather because the lake is very treacherous if it is stormy, but the day stayed clear and bright and hot and we made the trips safely without loss of life or gear. The worst problem was the sunburn on my forehead and nose. We camped near the school, about 400 yards from the site, and felt we had earned our excellent roast beef dinner that night.

Before long, we had named the boat *Mbota* (um-BOAT-uh), which was a pun on the English word *boat* and the word *(mbota)* that is the local name for Nile perch, a fish that grows to truly enormous size (hundreds of pounds). Lake Victoria is full of Nile perch that had been deliberately and misguidedly stocked in the lake to improve fishing and local nutrition. *Mbota* is a rather tubby, white fish with greasy, unappetizing flesh and carnivorous habits. Once in the lake, the *mbota* ate every other species of fish into near extinction and then started eating its own young. Now the fish ecology of the lake could be summed up as tiny *mbota,* small *mbota,* medium *mbota,* large *mbota,* and huge *mbota,* the last being much too large for any

The boat I am driving, nicknamed the *Mbota* (the Luo word for Nile perch), was our primary means of getting men and supplies from Rusinga to Mfangano Island. (© Mark Teaford.)

family to consume before the flesh spoils. The predominance of *mbota* created an enormous cottage industry in smoking fish, which was virtually the only way available to most people to preserve the flesh. The need to smoke fish in turn drove the people to cut down the trees on Mfangano for fuel, so that by 1986 the island that was once heavily wooded had only scarce stands of trees in the most inaccessible places. Rusinga had much the same problem.

Once Kamoya and I came upon a man on Rusinga who was busily hacking down a tree with a *panga*. Why, we asked him, was he cutting down all the trees? Didn't he know there would be no trees left for his grandchildren? With great dignity, he drew himself up and replied, "I am not cutting them all down. I am cutting them one by one." We had no reply.

Musa's site was now waist high in maize plants, so thickly covered that we could hardly find where the fossils had come from. The first thing we had to do was negotiate with the owner and his son about compensation. I couldn't buy the land outright—noncitizens cannot buy land in Kenya—so I had to buy the crop. When Kamoya and I started the discussion, we knew it was going to be a lengthy process and one in which we would need the chief. The next day, since we could not work on Musa's site until negotiations were finished, we took the boat around to look at the southern exposures where Louis and Mary had found insects, seeds, leaves, and the like. Everywhere we looked were maize and millet fields. There was practically no uncultivated land left on the entire island, though in Louis's day Mfangano had been almost uninhabited. Not any more: The population growth had completely outstripped the available land.

We went to see the chief but he was in Mbita and no one knew when he would return. Then we went to talk to the man who owned the *Proconsul* site. Somehow, though we had arranged to meet, he had gone over the small mountain in the middle of Mfangano to the other side of the island. Frustrated, we returned to camp and did small chores.

Next the son of the owner of the land upon which we were camped came to see us. He was a schoolteacher and spoke excellent English, but he was all pomp and posture. Obviously he was intent on getting the most out of us that he could for the privilege of camping for a few days on his father's land. Although we were indeed camped on his land, there was an identical piece of land immediately next to our campsite which was communally owned. I told him if he was troublesome we'd move our camp next door by 20 feet and he could explain to the D.O. why he didn't want to help us. He assured us he wanted to be helpful but demanded money for using his father's land. I told him that if he wanted to make out a bill to the Kenya government, he could, but I wouldn't pay him. After a long conversation, during which he drank two pots of tea and one of coffee, he left unpaid.

Our next visit was from the man who owned the maize *shamba* where Musa's site was. We sorted out the price by standing between the maize plants while the soil steamed from the recent rain and worked out how many and exactly which rows of grain we wanted to dig up. African-style, we did our sums by scratching the numbers lightly onto our skin with a stick. He asked for 3,000 shillings for a patch of scruffy plants that took up an area of about 20 feet by 30 feet in the field. We could have bought the land itself, had we been able to buy land legally, for much less. We counter-offered, suggesting that we would employ his two sons at the wheelbarrow and give him 1,800 shillings in compensation for the food he would lose. We'd also put a gate in his fence, so we didn't disturb his own gate and paths. The deal was agreed and we all shook hands on it.

Days later, the father of the schoolteacher turned up, irate. He claimed that the man with whom we had negotiated for the field was not the actual owner of the land, who was himself, but only his deplorably greedy nephew. We couldn't sort out who owned the land so we took the whole affair to the chief, who was finally back from Mbita. The chief in turn blamed the problem on us, saying we had paid too much money because we hadn't consulted him—but of course, we had tried to involve him and he wasn't on the island when we came to ask for his help. Eventually we worked out a reasonable (to us) but outrageously high compensation by local standards and we were able to return to work.

To our immense amusement, one of the most notable people we met on Mfangano was a hard-working, pleasant young man named Leakey. His name did not imply blood relationship but only that he had been born shortly after one of Louis and Mary's visits; the Luo often name their children after events associated with their birth. We hired him immediately and the Hominid Gang took special glee in ordering Leakey about by name. The joke was that Richard Leakey, then director of the National Museums of Kenya, was their boss and someone they could never order about in a peremptory fashion.

The really good finds of the 1986 field season started in late May. We had the usual trouble with rain and small rivers running through our tents and soaking everything, but eventually we started working Musa's site. On Saturday, May 24, Chris Beard found some vertebrae of *Proconsul nyanzae.* The first one had been chewed by a carnivore but the next was lovely: practically complete, with facets and the delicate accessory facets to show how it contacted the neighboring vertebrae in life. These looked like they came from the same individual as the vertebrae Musa had found when we visited Mfangano in 1984.

Chris, grinning broadly to himself, worked on his vertebrae while I excavated a *Proconsul* fibula that was in terrible shape, all splintered and fragmented, but nonetheless a fibula to go with the femur and ankle bones Musa had found in 1984. The best bone of the day was found by Iziah. He found a thin, flat piece of bone, then dug very carefully. He uncovered a left hipbone, or innominate, that looked somewhat like a chimpanzee's. It was riddled with plant roots and full of cracks, but it was almost perfectly complete. A large part of a pelvis is an incredibly important find if you want to reconstruct an animal's locomotion, since the pelvis incorporates the hip joints and the lowest joints in the spine. The specimen was very fragile and it was only the blindest luck that we had started to excavate before the plant roots completely destroyed it. I told Iziah I'd finish excavating it after lunch and he looked at me oddly, as if I were being lazy. I explained to him that it is never a good idea to work on delicate specimens in a hurry, when you are weary or hungry, or in the bright, flat light of midday. That was why I wanted to wait until later before removing the fossil from the sediments. Still, I was nervous that someone would come to see what we had been doing while we were away eating and would accidentally step on the pelvis, so I asked one of the local men working for us to stay there and guard it.

After lunch, we spent some time gluing together fragments from the sieving of the loose topsoil from Musa's site. From the splinters and bits, part of the left tibia, or shin bone, began to emerge, though the bone had been chewed at the knee end when it was fresh. Obviously this particular *Proconsul nyanzae* had been a tasty meal for some carnivore 18 million years earlier.

I stopped working to assess the pelvis again and decided not to try to clean it in the field. The Bedacryl was taking too long to dry in the moist heat, as usual, and I was afraid I wouldn't get the bone to ab-

From the foreground to the background: Chris Beard, Iziah Nengo, Mark Teaford, and Aila son of Derekitch working in the maize field on Mfangano Island. (© Alan Walker.)

sorb enough of the preservative before the end of work that day. Without being soaked in Bedacryl, and then dried, I knew the fossil would splinter into tiny pieces as I removed it from the soil. The sun sets abruptly near the equator; there is no long dusk or twilight as in higher latitudes, so I had to be finished with whatever excavation I was doing just before 6:00 P.M. The option of excavating by the uncertain light of a lantern and then walking home in the dark carrying a precious fossil was too foolish to contemplate.

Once I realized the situation, I stopped cleaning and started using nail scissors to cut through the maize roots that had penetrated and attached themselves to the bones. Once I had severed all the roots that might pull the fossil to pieces, I set about isolating the pelvis and its surrounding matrix on a fat pedestal of rock, and putting on a plaster jacket. I scribbled in my diary that night: "At least it's safe in camp tonight. It's a good feeling to have the first pelvis of any decent Miocene ape."

My wife, Pat, arrived at Mbita the next day, having driven up with her technician, Joan Fisher, to visit the camp for awhile. They were dusty and weary from the long drive over poor roads and she had pulled into the gas station in Mbita to try to find someone to check the brakes on the vehicle, which seemed to be failing. Kamoya discovered them at the gas station when he went to pick up gas for the *Mbota*. Though it was a rainy, cloudy day, he and I had motored across in the boat to pick Pat and Joan up and bring them back to the camp. These visits, even though they were brief, let her share in my research and fill me in on how hers was going. Besides, we missed each other terribly when we were separated by fieldwork, even though these separations were usual in our lives. A few days or a week together in the field was as good as a holiday on the beach to us.

Pat arrived in camp just in time to see the brand new, very old pelvis emerging from its matrix, and she was stunned. It looked a lot like a chimpanzee pelvis, with some important differences. And though many of the bones had been damaged by a carnivore, we had the makings of another partial skeleton: vertebrae, pelvis, femur, tibia, fibula, and ankle bones—only this time they were from *P. nyanzae*, not *P. africanus*. The museum number that was eventually assigned to the *P. nyanzae* skeleton was KNM-MW 13142. (KNM-MW is the National Museums of Kenya code for Kenya National Museum—Mfangano, which was formerly spelled Mwfangano.)

By the end of our work on Mfangano, we owned most of a defunct maize crop and we had found fossil riches. The final count was: six partial vertebrae, including part of the sacrum; a nearly complete left hipbone; most of the right femur and the shaft of the left femur; a fragmentary tibia and fibula; and the right talus and calcaneus (the two large bones of the ankle). The specimen was the best and most complete *Proconsul nyanzae* known.

A short while later, this partial skeleton provided a wonderful opportunity for Carol Ward, who was then my graduate student, to do an analysis that paralleled the epic 1959 monograph by John Napier and Peter Davis. It pleases me that the fascination and knowledge that John passed on to me, gained in part during his analysis of the *Proconsul*-in-a-tree from Rusinga, was transmitted in turn to Carol, who represented the next generation of scholars to take up the question of locomotion in Miocene apes. In addition to the maize-field skeleton from Mfangano, we had much of the left hand of *P. nyanzae* that I had found on our second day at Rusinga in 1985: the first hand skeleton known for this species. Without bragging, our team could fairly say we had contributed as much to the knowledge of *Proconsul nyanzae* as all previous finds combined.

The specimens from the Kaswanga Primate Site on Rusinga represented another quantum leap in evidence. By the end of the last field season in 1987, we had recovered much of at least nine individual *Proconsul africanus* skeletons and a single tooth of a tenth individual. Sorting them into individuals was a complicated task.

The immature individuals were easy to identify because they still had some of their milk teeth and the epiphyses of their long bones were unfused. By examining which milk or permanent teeth had erupted, and which epiphyses (if any) had fused, we were able to sort the nonadult bones into one infant, one juvenile, and four somewhat older subadult specimens of *Proconsul africanus*. We couldn't determine the sex of the infant and juvenile individuals because they were too young, but we could make some good guesses about the seven other *Proconsuls*. Among those seven more mature *Proconsuls* were two sizes of canine teeth, large and small. Among primates, males typically have larger, longer canine teeth and more robust bones than the females. If the same pattern holds true for *Proconsul*—and there is no reason it should not—then our collection included the two very young individuals of unknown sex, two juvenile females, two subadult males, and three fully adult females.

Many people get an odd intuition when they hear a recitation of the age and sex categories represented by these fossils, a sort of gut reaction that says *this is a family.* Probably this intuition is not true and certainly there is no way to prove that these individuals make up a family grouping. DNA cannot be extracted from such ancient bones with the techniques presently available. It is also difficult to conceive of a reasonable scenario by which a family of *Proconsuls* died at the same time and were buried and preserved so exquisitely. On the whole, I think it far more likely that we are looking at the remains of several carnivore dinners, collected individual by individual over time and brought back to a burrow. The rock in which the bones were found is indistinguishable from the surrounding rock, and the bones were concentrated in a burrow-shaped area. Bones in sheltered environments, such as caves or lairs, are often very well preserved as fossils.

Between Mfangano and Rusinga, we had an incredible number of bones of *Proconsul* and other Miocene creatures. The drawback to such an embarrassment of riches was that we had a truly enormous collection of bones to be cleaned, glued together, and allocated to the individual skeletons. Each individual skeleton, if complete, had over 200 bones. Hand and foot bones were particularly troublesome. Each whole primate skeleton has 28 finger phalanges and another 28 toe phalanges, for a total of 56. Metapodials, the bones that underlie the palm of the hand or the sole of the foot, are long and slender like most phalanges; each individual has 20 metapodials. Carpals and tarsals are the small bones of the hands and feet; each individual has 18 carpals (one more than humans, as it happens) and 10 tarsals. Thus there are 104 bones in the hands and feet of a complete skeleton. With at least nine individuals at Kaswanga, we could expect to find as many as 504 phalanges, 180 metapodials, and 252 carpals and tarsals, or 936 bones to be separated from each other and clustered into their original hands and feet.

We never expected to find them all and we didn't, since some bones had probably washed away or been destroyed by weathering, goats' hooves, or plant roots before we got there. Of course, the carnivores that dragged the *Proconsul*s into the tree hollow might have lost some pieces along the way, too. We did find 658 hand or foot bones, including incomplete fragments that may glue together, and that was more than enough. Simply recording and numbering the specimens was a daunting job; matching them into whole bones was downright eye-straining. One of my postdoctoral students in those years was David Begun, who became an expert in the morphology (or shape) of hand and foot phalanges in primates and what shape revealed about function. David embarked on the truly heroic job of sorting, gluing, measuring, and analyzing these remains as happily as if I'd handed him a chest full of jewels. In the years since then, he has become a professor at the University of Toronto and a recognized expert on Miocene hominoids from Africa, Hungary, Germany, Spain, and China, among other places.

A few weeks of uncanny luck in the museum, followed by four field seasons of excavation and exploration, had yielded fossil riches no one ever expected. We had vastly increased the sample of Miocene hominoids, the group that surely gave rise to modern apes and humans as well as many extinct forms. We had found many skeletal elements of *Proconsul africanus* and *Proconsul nyanzae* that had never before been known. And the extraordinary quality of the fossils—the whole anatomical segments of skeletons—was going to make new kinds of analyses possible. It felt as if we had discovered a new continent, full of surprises and opportunities, and maybe a few unanticipated pitfalls.

7

How Did It Move?

As the scientist in charge of the expeditions that recovered the abundant new *Proconsul* material, I had the right to decide who worked on what. My main intent was to give as many opportunities to up-and-coming young people as I could, to get their careers off to a good start. Sometimes it felt as if every paleoanthropologist I knew—and some I didn't—were asking to study the new material. The sensation was, I imagine, a little like being the man in charge of registering land claims at the beginning of a gold rush. I made sure that those graduate students and postdoctoral fellows at Hopkins who were interested in the material had good projects to work on. There were still plenty of fossils left over for Mark and me to research in collaboration with various specialists.

The easiest way for me to describe the research discoveries that followed is to cluster them into broad topics, ignoring strict chronology. There was a flurry of work, much of it simultaneous, and often one person's ideas or findings influenced another's. That sort of rich interactive environment fosters new ideas and good research, I find. Working in a group, even if the participants are scattered and can't talk every day, makes your mind buzz with ideas and possibilities in a marvelous way.

One of the most important questions that needed to be addressed was: How did *Proconsul* move? From the very outset, *Proconsul* had figured prominently in locomotor hypotheses. How an animal moves has a strong influence on many different aspects of its ecology: what sorts of food it can obtain, where it can raise its young, where it is safe to sleep, how large its territory is, and, sometimes, how fast its babies grow.

The additional parts of the *Proconsul*-in-a-tree skeleton had confirmed much of what John Napier and Peter Davis had surmised years earlier, with a few important exceptions. That specimen made clear how unexpectedly robust the bones of this small *Proconsul* were, given its small body size. With a more complete humerus, the emphasis John and Peter gave to the few supposed brachiating adaptations of *Proconsul* seemed unwarranted; the best that could be said was that *Proconsul* had adaptations to a generalized, arboreal lifestyle and was capable of a wide range of movements at all of the limb joints. Of course those movements probably included some suspension, but *Proconsul* was no specialized brachiator.

Because the new discoveries included the lower half of the skeleton of *Proconsul nyanzae,* they offered a chance to have an independent look at the locomotor adaptations in a larger-bodied member of the genus. As a graduate student, Carol Ward took full advantage of the opportunity.

Carol was tremendous fun to have in my lab because she was bright and full of enthusiasm. For a while, she had a very curly hairdo; there was something about her big, eager blue eyes, freckled face, and the waves of curls floating joyously out from her face that seemed to express her personality. During her first year or so of graduate school, Carol seemed a little vague: very smart but distracted, not knowing what she wanted to do. This may have been because she was at an early stage in her training or it may have been related to her long-distance romance, for during that period she became engaged to the anthropologist Mark Flinn, now her husband and colleague at the University of Missouri. During Carol's graduate school years, Mark was employed at universities far away from Baltimore, so they didn't get to see each other very much. Sometimes I used to find her playing games on the computer in the lab, until I finally told her to stop wasting her time and her mind. It was one of the few times I can remember ever needing to be gruff with Carol. I don't know what exactly precipitated the change, but one day Carol suddenly knew exactly what she wanted to do. She designed and carried out her research and finished her thesis, earning her degree in 4 and 1/2 years flat—the fastest any student in that program ever did. Watching her develop as a professional was like looking through a camera lens. At first, everything was fuzzy and I couldn't quite make out what I was seeing. Abruptly and all at once, the picture was in exquisitely clear focus. Ever since then, Carol has been a leading scientist in the study of fossil primate limb bones and what they imply about locomotion.

One of her first publications was a brief and elegant paper she coauthored with Mark Teaford and me that was entitled, evocatively, "*Proconsul* Did Not Have a Tail." It concerned a single specimen from the Kaswanga Primate Site, a bone that was part of the juvenile skeleton from the tree. The specimen is the last sacral vertebra.

In a fully grown animal, all of the sacral vertebrae (the total number of them varies from about three to seven in different primate species) are fused together to form the bony sacrum. The sacrum lies between the innominates, or hipbones, and together the three bones form the bony pelvis. At the bottom of the last sacral vertebra, tailless primates like humans have a few vestigial vertebrae known as the coccyx. In tailed primates, the bottom of the last sacral vertebra articulates with the first caudal, or tail, vertebra. Because the specimen came from such a young animal, the last sacral vertebra was not yet fused by bony tissue to the others, which were never found. All we had, then, was this small piece of the sacrum, but it revealed its secrets.

Like other vertebrae, a sacral vertebra has a weight-bearing, roughly cylindrical portion called the body, various projections by which it articulates with adjacent vertebrae, and a bony arch that loops backward from the body. In the vertebrae of the neck and trunk, that bony arch houses and protects the spinal cord, so it is known as the neural canal. The first few sacral vertebrae also have a neural arch.

In tailless animals like humans, by the level of the fourth lumbar vertebra, the thick spinal cord has diminished to a stringy collection of nerves known as the cauda equina, or "horse's tail," because of its appearance. Thus the sacral vertebrae do not enclose a single spinal cord at all, but only enclose a bundle of nerves. These anatomical facts explain why, when physicians need to sample the meningeal fluid that bathes the brain and spinal cord, they do a spinal tap by sticking a needle between the fourth and fifth lumbar vertebrae. (I can say from personal experience that having this done can be a most painful experience.) At this point on the spine, there is no solid spinal cord but only the cauda equina, which is bathed in spinal fluid that is in turn contained by membranes, or meninges. As the needle intrudes through the meninges, the cauda equina is pushed away from the needle by the fluid, so there is little chance of puncturing the nerves. Because in tailless animals there are no caudal vertebrae and no tail to innervate, the neural canal diminishes fairly sharply from the top to the bottom of the sacrum. The neural canal is very small or nonexistent at the bottom of the last sacral vertebra.

In contrast, in tailed animals, the spinal cord continues farther down in the spine and sometimes passes through the neural canals of the sacral vertebrae. The width of the neural arch remains fairly constant from the top to the bottom of the last sacral vertebra because the spinal nerves still have a substantial distance to run in order to innervate the tail to its end.

For the same reason, the diameter of the body of the sacral vertebrae diminishes slowly in tailed animals and rapidly in tailless ones. Viewed from the back, the diminution of the vertebral body can be measured as an angle or degree of tapering. If the angle of tapering or wedging is positive, then the neural canal is decreasing in size, as in tailless animals; if it is neutral or zero, the width of the neural canal is remaining stable; and if the angle of wedging is negative, the neural canal is increasing in size.

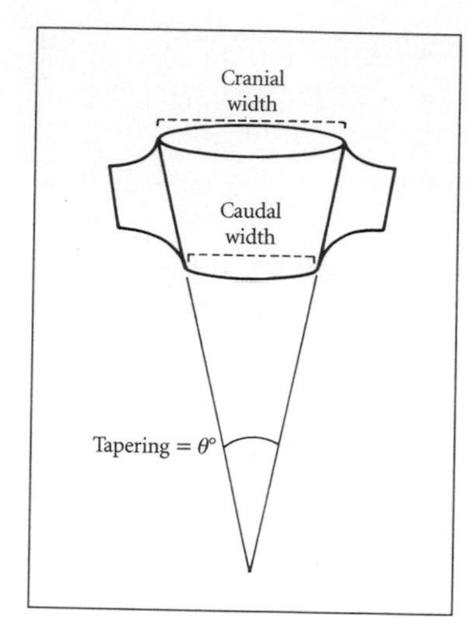

This diagram shows the last sacral vertebra of a tailless primate. In *Proconsul* and other tailless creatures, the sacral vertebra tapers sharply and has a positive tapering or wedging angle θ. Tailed animals have negative or neutral wedging angles. (Reproduced with the permission of Elsevier from C. V. Ward, A. Walker, and M. Teaford, "*Proconsul* Did Not Have a Tail," *Journal of Human Evolution* 21 [1991]: 215–220.)

Carol measured the degree of wedging in the last sacral vertebra of eighty-five individual animals from nine species with varying tail lengths. The long-tailed monkeys she examined (spider monkeys, vervet monkeys, and colobus monkeys) had a mean wedging angle of -11 degrees. That meant that the neural canal was increasing in diameter at this point, because the tails of these species are long and heavily innervated. Among the short-tailed monkeys in her sample (baboons and macaques), the average degree of wedging was +4.6 degrees. Neither baboons nor macaques use their tails much, so the nerve tissue is less extensive than in longer-tailed monkeys. Finally, she measured the vertebrae of tailless apes (chimpanzees, gibbons, gorillas, and orangutans) and derived an average degree of wedging of +30.8 degrees. In these animals, the spinal cord comes to an abrupt end shortly below the last sacral vertebra.

The *Proconsul* vertebrae had a wedging angle of +29.3 degrees. No tailed primate in Carol's sample showed such a strong degree of wedging while every tailless primate did. Like humans and living apes, *Proconsul* probably had a small coccyx but no externally visible tail.

In the paper, Carol deadpanned: "Tail loss is a diagnostic feature of hominoids." In other words, the fact that apes have no tails is a key feature that defines them as apes. The same is true of humans, which is part of what makes them hominoids along with apes. The corollary of "*Proconsul* did not have a tail" is "*Proconsul* is an ape," and Carol had proved it.

Before concluding the paper, we made one final observation. Whatever the cause or adaptive advantage of taillessness was to apes, the evolutionary loss of a tail could not be construed as a by-product of a brachiating, suspensory form of locomotion, because *Proconsul* had precious few adaptations to brachiation in its limb skeleton.

Later, Terry Harrison of New York University published a different interpretation of the same fossil Carol, Mark, and I analyzed in the "tailless" paper. He thought I had misidentified the specimen as a last, or terminal, sacral vertebra and suggested it might be a caudal, or tail, vertebra, which would mean *Proconsul* did have a tail. Harrison also felt I had misidentified two other vertebrae as lumbars from the lower back when they, too, were tail vertebrae. To be sure, the specimens were broken and distorted and difficult to interpret. If Harrison was right, then one of the key features indicating that *Proconsul* had already specialized and become a stem ape—an early member of the lineage that culminated in modern apes and humans—would disappear.

The matter was controversial until very recently. Then Carol, Mark, and I were invited by Masato Nakatsukasa and Naomichi Ogihara of Kyoto University to review the *Proconsul* vertebrae in light of new fossil vertebrae from Kenya.

The Japanese team, led by Hidemi Ishida of Kyoto University, had discovered (among other body parts) some vertebrae of another Miocene ape, *Nacholapithecus kerioi*. *Nacholapithecus* comes from a 15-million-year-old site in Samburu, northern Kenya, so it was a more recent (less ancient) early ape than *Proconsul*. The fossils were first described in a paper in July of 1999, when the team published the results of a series of expeditions to the Samburu Hills of Kenya.

The Samburu fossils had initially been attributed to *Kenyapithecus africanus*, an already-known Miocene ape species, but the discovery and analysis of a partial skeleton prompted a reclassification that resulted in the erection of a new genus and species, *Nacholapithecus kerioi*. Nachola is a town in Kenya and *pithecus* indicates the species is an ape.

The bones of the *Nacholapithecus* partial skeleton are unfortunately badly fragmented and somewhat distorted, meaning that in many respects they are difficult to interpret with confidence. As best as can be judged from the specimens, the Samburu skeleton had long-toed feet and seemed to have an unusually long forelimb compared to its hind limb. Long toes are useful for grasping branches and long forelimbs are often seen as an indication of a brachiating habit; both suggest that this species may have been arboreal in a different way from *Proconsul*. The shape of the additional fossils from the forelimb supports the idea that *Nacholapithecus* was an arboreal animal that used its arms a great deal in locomotion, though full-blown brachiation, such as is practiced by gibbons, is unlikely.

Among the skeletal remains are nineteen partial and largely complete vertebrae from the skeleton and among them was one vertebra from the coccyx of *Nacholapithecus*. Since the coccyx is the vestigial remnant of a tail in tailless animals, this specimen convinced Nakatsukasa and his colleagues that *Nacholapithecus* lacked a tail. The question then became: Had taillessness begun earlier, with *Proconsul*, or not? Was taillessness a fundamental attribute of the earliest apes or had it evolved later?

The team took CT scans of many modern primate vertebrae and the two vertebrae from *Proconsul* that Carol and I had identified as lumbars and that Terry had identified as caudals. Both fossil specimens had a complex internal structure with a fine trabecular meshwork of bone. Modern lumbar vertebrae consistently have such a structure and caudal vertebrae do not, so this observation—along with some of the details of the morphology of the fossils—resolved the debate about the *Proconsul* vertebrae: They come from the lumbar region, as I had thought.

The most controversial specimen was the one Carol had identified as the last sacral vertebra. Because one *Nacholapithecus* specimen was a coccygeal vertebra, which in life is directly next to the last sacral vertebra, it was important to compare the *Proconsul* and *Nacholapithecus* specimens. In a number of detailed features, the *Proconsul* specimen differed from the *Nacholapithecus* specimen and resembled sacral vertebrae from modern primates. Neither closely resembled caudal vertebrae of tailed primates. Thus although the *Proconsul* specimen was distorted and broken and difficult to interpret, we were reassured that this specimen is a sacral vertebra and that *Proconsul* cannot have borne a tail. We also felt more comfortable in our conclusion that *Proconsul* was an early stem ape lying near the beginning of the ape lineage which later included *Nacholapithecus*—and that such stem apes seemed to be tailless.

For her Ph.D. thesis, Carol did a series of remarkable studies on the partial skeleton of *Proconsul nyanzae.* Before our discoveries in the 1980s, the postcranial (literally "behind-the-head") bones of this species were hardly known, so Carol had wide open intellectual territory ahead of her. In research such as this, one of the first obligations of any scholar is to write a thorough description of the anatomy of each of the specimens. In this case, the fossils comprised: six vertebrae from the lower back; the sacrum (in two pieces); the left hipbone; the two femora, or thighbones, one of which was almost complete; the tibia, which had been badly damaged by the farmer's hoe; the fibula, which was intact at the ends though most of the shaft had been shattered by the roots of the maize plants; and the complete talus and calcaneus, the two large bones of the ankle.

There was an interesting discussion among the members of Carol's thesis committee about the value of publishing anatomical descriptions and whether she should include them in her thesis. A technical description of a fossil intentionally omits analysis or discussion, apart from specifying ways in which the new fossil can be distinguished from previously known ones, so descriptions may seem rather mindless or mundane. Descriptions are often boring to read—until you need a piece of information that they contain, and then every word is riveting. While the interpretations may come and go as theories and approaches change over the years, the basic description of a fossil should endure forever. Some members of Carol's committee were not paleontologists and had other specialties in functional anatomy. This meant that they hadn't ever written or used an anatomical description in their own work and had not considered how important such dry descriptions were. They urged her to cut them out of her thesis completely. For Carol's sake as a scientist, and to preserve the lasting value of her work, I had to intervene and explain to them why Carol needed to include lengthy and detailed descriptions in her thesis.

Writing a thorough, clear anatomical description of every bump, every groove, and every feature on a bone is an endeavor that requires a lot of skill, a systematic mind, and a good deal of detailed knowledge. Assigning a bright student the task of describing new fossils is a terrific way to make him or her observe, think, and write precisely; it is a crucial element in his or her professional training. The process of writing the description also commits the shape of a bone or fossil to the student's memory forever.

Think of having to describe, say, a Queen Anne table in such a way that someone who has never seen one could read your description and decide whether or not the object sitting in front of him or her was a Queen Anne table. You would have to be tremendously knowledgeable about furniture and how it was constructed and decorated to write such a description; you might have to specify in detail how Queen Anne tables compared to Louis XIV tables or Arts and Crafts period tables or Georgian tables, and so on; and, once done with all that, you'd probably never mistake anything else for a Queen Anne table. So it is with fossil descriptions.

What's more, this basic and first description of a specimen affords other paleontologists the highly detailed and technical information that they may need for their own studies. Such descriptions are a way of making the basic, uninterpreted data available to others. Since no one had ever seen these skeletal elements from *Proconsul nyanzae* before, the descriptions were very important indeed.

The real point of Carol's work was the functional analysis. She wanted to deduce function from these body parts and find out how *Proconsul nyanzae* used its back and lower limbs to move around the world. Another issue was, of course, whether or not *P. nyanzae* had the same functional anatomy as its smaller relative, whose skeleton had been found in the fossil tree on Rusinga.

One of the first interesting facts Carol discovered was that *Proconsul nyanzae* had a most peculiar back. Living great apes have fairly rigid lower backs compared to those of monkeys and this is reflected in both the number and shape of their vertebrae. Great apes have the least flexibility in their lower backs of any group of higher primates. They have only three or four lumbar vertebrae and the movement between each adjacent pair of vertebrae is physically constricted in several ways. On a great ape, the body of each lumbar vertebra is roughly constant in height from the ventral, or stomach, side of the vertebral body to the back. In the living animal, the vertebral bodies are arranged in a continuous column and each is separated from its neighbor only by an intervertebral disk. Thus the constancy in height of each vertebral body means that the vertebral column is almost straight in the lower back and does not naturally bend much.

Each lumbar vertebra also has various bony protuberances, or processes, that perform two different functions. First, the processes of one vertebra articulate or interlock with those from the next vertebra, which insures that the entire vertebral column works as an integrated unit. Second, these processes serve as the attachment sites for back muscles. In great apes, the possible movement between each pair of vertebrae is relatively slight because of the shape of the articular, or vertebral, processes. The size and placement of these processes show that the muscles of the lower back are relatively small. Thus not

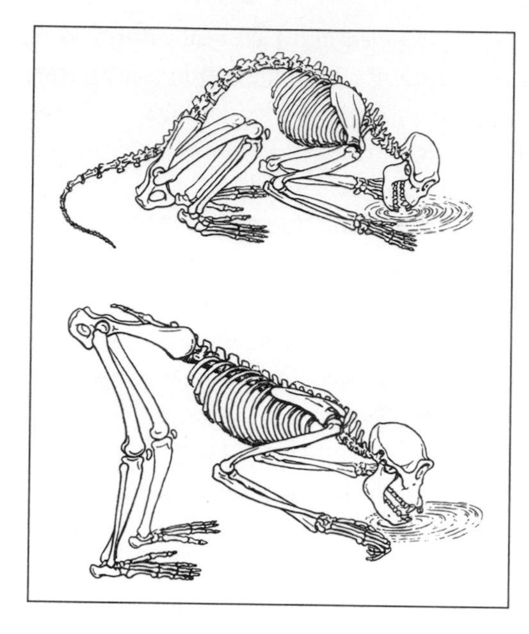

A monkey (top) bends its long flexible spine, which has six to seven lumbar vertebrae, in order to drink. An ape (bottom) has fewer lumbar vertebrae and less flexibility between each adjacent pair, so it must bend at the hip to drink from a pool. (From A. H. Schultz, *The Life of Primates* [London: Weidenfeld and Nicolson, 1969], p. 68.; © A. H. Schultz-Foundation.)

only is the vertebral column straight as a function of the shape of the bodies, but there is only a minimal amount of movement possible between each pair of vertebrae in response to muscular action.

Gibbons and siamangs, the lesser apes, have five or six vertebrae in the lower back and somewhat freer movement, as is shown by the fact that their vertebral bodies are shorter at the front than they are at the back. Thus the lower part of a gibbon's back is naturally a little convex seen from behind and its vertebral column is capable of greater rounding through muscular action, a movement that is hardly possible for a chimpanzee or gorilla.

Monkeys of various species show the greatest flexibility in the lower back. They have six to seven lumbar vertebrae and an even more marked difference in the height of the vertebral body from front to back. The vertebral processes are robust, to anchor strong back muscles, and these processes are positioned to enhance (not restrict) the possibility of movement between each pair of vertebrae.

These seemingly unimportant differences in the shapes of the lumbar vertebrae in great apes, lesser apes, and monkeys have a dramatic effect on the movements and postures that can be performed by these animals. They are part of what makes an ape look like an ape or a monkey like a monkey, even from a distance. For example, if monkeys and apes need to drink from a pond, they do it in different postures. Flexible monkeys squat on the ground and bend over, curving their spines to reach their lips to a pond to drink. Stiff-backed apes keep their legs much straighter, bending deeply at the hip to lower their torsos straight-backed until their heads reach the water. There are similarly striking differences in the actions that can be used to propel animals of different shape across the ground or through the trees.

Measurements of the *Proconsul nyanzae* vertebrae left no doubt that it had at least six lumbar vertebrae (even though we found only five of them) and had a long, curved, and flexible lower back. The robust processes from each vertebra are placed in a monkeylike position in *P. nyanzae,* allowing considerable movement of the lower back that was powered by strong muscles. This same pattern can be observed in *Proconsul africanus,* too, which means that both fossil species had long flexible backs, unlike all living apes but like living monkeys.

The hipbone of *Proconsul nyanzae* is also monkeylike in many ways. Though Carol didn't have both hipbones to work with, she did have one that was almost complete and she could "mirror-image" it: draw and/or model the hipbone of the right side as if it were identical to the left hipbone, only reversed as she would see it in a mirror. By putting the real hipbone together with its mirror image, and then reuniting the two sacral pieces into one, Carol could reconstruct the overall shape of the pelvis fairly accurately. She could also see how the pelvis attached to the lower back, because she had lumbar vertebrae. The result confirmed that *P. nyanzae* had both a long, flexible back and a deep, narrow torso. This is the shape typical of many modern quadrupeds, such as dogs, cats, sheep, horses, or cattle, as well as monkeys. For a primate, the monkeylike torso shape is primitive or ancestral, which reveals another way in which this early ape was still primitive. Modern apes have torsos of quite a different shape—broad, short, and shallow front to back, much like human torsos—with a more rigid lower back. Also like humans, modern apes spend a lot of time in postures or locomotion in which their trunks are upright, even though they are not walking on two legs, as we do, but hanging from two arms. Apparently *Proconsul* habitually went on all fours rather than hanging from its arms or walking on its legs.

Because *Proconsul* seemed so monkeylike in the shape of its torso and the arrangement of its back and pelvis, Carol looked for clues about another feature. Some higher primates have hairless, leathery sitting pads known technically as ischial callosities. This sitting pad is an external structure underlain by a part of the pelvis known as the ischium. To support and anchor the thickened skin of the sitting pad, each ischium has a tuberosity, or "bump," that is large, flat, and expanded.

Among the higher primates of the Old World, there is a gradient in the appearance of the various features that make up the sitting pad and its support. Monkeys have large and often obvious sitting pads with a distinctively shaped ischial tuberosity underlying them; gibbons have small sitting pads, which are farther toward the back of the ischium than monkeys' sitting pads, as is the expanded and flattened ischial tuberosity; and great apes lack the sitting pad entirely and have an ischial tuberosity that is not expanded or flattened.

Frustratingly, most of the relevant part of the ischium is missing in *Proconsul nyanzae;* the fossil is broken away just short of where the expansion of the ischial tuberosity would begin, if it were present. Carol studied the part of the ischium that was preserved and realized that, in animals with expanded ischial tuberosities and sitting pads, the cross-section of the ischium above the tuberosity was always a flattened oval because the surface of the bone flared to meet the oval tuberosity. Instead, in *Proconsul* the ischium had a rounded cross-section with no sign of flaring. In other words, what was preserved wasn't the right shape for an expanded ischial tuberosity to have been present. The closest match for the ischial portion of the pelvis of *Proconsul nyanzae* is a chimpanzee's ischium, not a baboon's. Once again, a detailed study had shown that *Proconsul* was a mosaic of monkeylike features (long, flexible back; strong back muscles; deep, narrow torso) and apelike ones (no ischial expansion and no sitting pads; no tail).

As Carol moved down the partial skeleton to analyze the hip joint, she encountered additional anatomical features that could be considered either apelike or monkeylike, but only if the monkey in question was a New World monkey. While Old World monkeys with arboreal habits tend to specialize in rapid movements—leaps, jumps, and runs—New World monkeys use more varied and often slower movements in the trees. These differences in locomotor habits among monkeys are reflected closely in the anatomy of the hip, thigh, knee, and ankle.

The key to efficient, rapid movements and sharp acceleration is to perform a powerful, coordinated thrust by straightening the hip, knee, and ankle simultaneously. Jumping animals all do this. The way to maximize the power of that thrust is to restrict movement at the hip, knee, and ankle to a fore-and-aft plane so that all of the muscular force is expended in a single direction. In Old World monkeys—or, for that matter, in other animals extremely well adapted for running, such as horses—the shape of the head of the femur is ovoid, not spherical, and the hip socket itself is deeper, to guard against dislocation. These structures help keep the legs pulled in underneath the body so that the animal can move straight forward. These special adaptations are seen most clearly in ground-dwelling animals, because running on a flattish substrate poses different demands from running on curved branches. Indeed, to judge from the earliest such fossils that are known, Old World monkeys had a period in their ancestry when they were partly or largely terrestrial.

Another set of fossils included in Carol's tour of the lower half of *Proconsul* were fossil kneecaps from Rusinga, which she analyzed in collaboration with several of the team members, me included. Four kneecaps, or patellas, were recovered from the Kaswanga Primate Site and were of a size to belong to the little *Proconsul* species. Another two patellas, of similar shape but distinctly larger size, were found at R106 and are of the right size and shape to fit with the partial skeleton of *P. nyanzae.*

Once again, the different ways that monkeys and apes use their hind limbs is reflected in the shape of the kneecap. The kneecap is part of a pulley system involving the big quadriceps muscle at the front of the thigh. That muscle originates high on the hipbone and thigh and narrows down to a thick, strong tendon at the knee. This tendon inserts on the kneecap, which rides up and down in the patellar groove on the distal (or knee end) of the front of the thighbone. A continuation of the quadriceps tendon then runs over the surface of the kneecap to insert on the front of the tibia, the major bone of the lower leg. When the quadriceps muscle is contracted, the first tendon pulls on the kneecap, which causes the second tendon to pull on the tibia. The result is a straightened knee as the kneecap moves up its groove. When the quadriceps muscle relaxes, the tendons relax, the kneecap slides down the groove a little, and the knee bends.

The kneecap of a fast-moving Old World monkey is deep and narrow and fits tightly into the high-sided patellar groove. Animals with even stronger adaptations to running and rapid acceleration, such as horses or antelopes, show similar restrictions to an even greater degree, so that the animals can direct all the power of their muscles into a rapid movement. The tight fit and deep groove are designed to keep the kneecap from dislocating when the muscular pull is sudden and strong.

My wife, who once dislocated her kneecap, tells me that it is a very effective way to stop all desire for further movement for the immediate future. A dislocated patella would disable an animal to the point of putting it in real danger.

Apes and New World monkeys, adapted for a wider range of

movements, have wider and shallower kneecaps that slide along in a broader, less restricted patellar groove, permitting a wider range of motion at the knee. As she might have predicted, Carol and her colleagues found that both species of *Proconsul* had apelike kneecaps that attested to a slow and deliberate flexion and extension of the knee.

In animals that run, jump or leap, the motion of the ankle joint is similarly restricted to a single plane by the anatomy of the joint. Often animals that habitually run or leap have a compromised grasping ability because their fingers and toes are short, presumably to keep them from being injured. Short digits facilitate a single, sharp push-off movement, whereas long ones are better for grasping branches or small supports. Again, the extreme case in this regard is the horse, in which all of the digits have been lost except one and the grasping ability of the foot and hand are completely sacrificed in favor of tough, strong hooves.

How did *Proconsul* match up to these different sets of adaptations? In both *Proconsul* species, the hip and thighbone were built for a wide range of movements in many directions. The head of the femur—the ball of the ball-and-socket joint at the hip—was nearly spherical, so that the leg could be twisted and rotated into many different positions. Like the shape of the hip joint, the shape of the knee joint and the kneecap permitted a wider, less restricted range of motion than is possible in committed runners or leapers. The ankle joint itself was also very flexible and the lower leg and ankle bones of *Proconsul* revealed markings for the powerful tendons and big muscles that worked the foot, especially the grasping big toe.

The total picture Carol constructed from her careful studies was very similar to the picture of *Proconsul* locomotion that we had deduced from examining the bones of the smaller species. She was able to address some new issues, like the pelvic shape, because she was working with a skeleton that included some body parts still unknown for the smaller *Proconsul*-in-a-tree. Except for its primitive, monkeylike torso shape, much of the skeleton of *Proconsul nyanzae* was adapted to habitual, careful movement through the trees, grasping with hands and feet, and using limbs in many different positions instead of specializing in one. Both partial skeletons of *Proconsul* suggested that these animals clambered along in the trees in a quadrupedal fashion. Though few of the specialized adaptations associated with hanging by the arms were found, *Proconsul*'s body revealed its hominoid status in its lack of a tail and sitting pads. Thus *Proconsul* was an ape—a stem ape—but not one with the form of locomotion that is dominant among the modern apes.

The small- and medium-sized *Proconsuls* differed from each other primarily in size, not in shape or function. This spurred Carol to take another look at the only two vertebrae attributed to the largest species, *Proconsul major.* These had been found at Moroto, Uganda, by my former colleague in Uganda Bill Bishop and had been described by Michael Rose and me in 1968. From our paper, Carol knew these fossils were lumbar vertebrae similar to those of a great ape, but she was a little surprised to see how very different they were from other *Proconsul* vertebrae. The bodies were similar in height at back and front and the processes were shaped and placed to enhance the rigidity of the lower back, which implies that *Proconsul major* had a shorter, less mobile back than the other *Proconsul* species and quite possibly a broad, shallow torso, unlike the other *Proconsuls*. If anyone knew the vertebral anatomy of *Proconsul,* it was Carol, and these specimens just didn't fit with the other specimens.

After completing her study, she endorsed a suggestion that Meave and Richard Leakey and I had made: that the Moroto vertebrae should be removed from the genus *Proconsul* altogether. We wanted to reclassify these specimens as another new and large-bodied ape we had found, *Afropithecus,* a genus that hadn't been known at the time that the Moroto fossils were found.

8

How Many *Proconsuls?*

The abundance of specimens meant that the time had come to take a good hard look at *Proconsul* and the differences and similarities among the species within that genus. If the Moroto vertebrae wasn't *Proconsul major*, what was *P. major?* Was *P. nyanzae* really just a very large version of the smaller *Proconsul?* Were all the specimens attributed to *Proconsul* one species, or several?

To answer these questions meant going back and reviewing exactly what criteria had been used to define the genus *Proconsul* and each of the species in it. To begin at the beginning, *Proconsul africanus* was first named by Arthur Hopwood in 1933, on the basis of the left maxilla from Koru. That specimen was the type used by Hopwood to define both the genus and the species, so reexamining the anatomy of the Koru jaw and teeth was key.

To understand the importance of reexamining the Koru jaw, you need to know some of the detailed rules about nomenclature, the formal naming of specimens. When a species is first defined, the person who proposes the new name must identify a holotype—the type specimen—that exemplifies the features of the new species. The formal species name is thereafter attached to that particular specimen. (The naming of a new genus works the same way.) If, for example, the holotype is initially grouped with other specimens and is later removed from the group, the name goes with the holotype, not the group. What this means is that the Koru specimen was literally the embodiment of *Proconsul africanus* and anything that was classified as the genus *Proconsul* or the species *P. africanus* needed to match that particular specimen in its diagnostic details.

In 1950 and 1951, Le Gros Clark and Louis Leakey suggested that specimens from Rusinga represented *Proconsul africanus* and a larger, related species they called *Proconsul nyanzae,* while material from a still larger hominoid from Songhor, and a few fragments from Rusinga, were *Proconsul major.* Le Gros and Louis were handicapped by the fact that they had no radiometric dates from any sites, since the necessary techniques had not yet been developed and applied to fossil sites, so they could not consider the relative antiquity of the fossils except to say that they all were Miocene. Le Gros also had a natural tendency to lump specimens into only a few species, while Louis was a natural splitter who elevated small differences into sometimes unwarranted species differences. In 1968, Mike Rose and I classified the Moroto specimens as *P. major.*

In 1965, Elwyn Simons and David Pilbeam, then both at Yale University, reviewed the entire record of fossil apes. In a frenzy of taxonomic lumping, they suggested that *Proconsul* was actually a subgrouping within the European genus *Dryopithecus* and belonged in the subfamily Dryopithecinae. Though Elwyn and David acknowledged the distinctness of the African fossil apes called *Proconsul,* for quite a few years after their publication the name *Proconsul* was out of fashion. That is why some publications from the late 1960s and 1970s confusingly use the term *Dryopithecus africanus.* Eventually the name *Proconsul* came once again to be accepted as the appropriate generic name.

A few years later, in his Ph.D. thesis, David speculated that *P. major* was ancestral to gorillas, *P. nyanzae* to chimpanzees, and *P. africanus* to humans, repeating a suggestion made by Louis earlier. As more and more fossils accumulated, not only the relationships among species but the very way the specimens were sorted into individual species came into question.

The uncertainty arose in part because of preservation issues. Generally, teeth and lower jaws are among the most common body parts preserved of any fossil animal. Teeth and jaws are very dense, which helps them survive all kinds of destructive agents and makes them unattractive to carnivores after the jaw muscles have been eaten. As might be expected, most of the specimens of *Proconsul* were dental. There are particular anatomical details about *Proconsul* teeth that let a paleontologist recognize them as *Proconsul* without too much difficulty. When we had relatively few dental specimens, the teeth clustered fairly neatly into three size groups—small, medium, and large—which were taken to mean *P. africanus, P. nyanzae,* and *P. major,* respectively. Within each cluster, everyone expected to find slightly larger (male) and slightly smaller (female) specimens.

With time and additional specimens, the simplicity of this scheme splintered like a bone crunched up by a hyena. In 1972, Leonard Greenfield noted that the majority of the canine teeth of *Proconsul africanus* appeared to be from female individuals. Was this an odd happenstance or did it mean something more important about our classification methodology? In 1981, a clever Australian researcher, Wendy Bosler, went a step further, noting that while the *Proconsul africanus* canines were all female, the *Proconsul nyanzae* canine teeth were predominantly male. Why, she asked, with such an abundance of fossils from Rusinga and Mfangano, were males of *P. africanus* unknown? That pertinent observation led my colleague Jay Kelley, then a hard-working graduate student at Yale, to ask if specimens that were being called *Proconsul nyanzae* were nothing more than misidentified males of *Proconsul africanus.* Might the two "species" simply be sexes?

Jay's suggestion made everyone stop cold and think hard. In time, objections to his interpretation arose, the first of which relies on the principle of uniformitarianism, which can be simply stated as: *The present is the key to the past.* What this axiom means is not that the present is *the same as* the past; we know that evolution has occurred and continues to occur, so obviously species change over time. What the principle of uniformitarianism asserts is that *the processes at work now*—the biological, geological, chemical, and other "rules"—*were also at work in the past.* Gravity was not suspended during the Miocene, nor were fundamental rules of metabolism, biomechanics, fossilization, preservation, embryonic development, and so on.

Where uniformitarianism comes into play in this case is in the degree of sexual dimorphism—the anatomical differences (other than genital ones) that distinguish males from females—seen in higher primates. Some living primates show a high degree of sexual dimorphism; an oft-noted feature is that males' canines are longer and stouter than females'. Even in the cheek teeth (the molars and premolars), male primates usually have bigger teeth than females. Orangutans are the most dimorphic living hominoids in this regard, with a male having cheek teeth the combined area of which are, on average, 1.22 times larger than a female's combined area. Another dimorphic feature is that the crests and ridges on the bones for muscle attachments are usually larger and more pronounced in male higher primates than in females. Most strikingly, males may weigh twice as much as females. For example, a male gorilla commonly weighs about 300 to 350 pounds (140–160 kilograms) while the female weighs only 154–220 pounds (70–100 kilograms).

The range of sexual dimorphism in higher primates is well documented, especially as regards tooth size. The fundamental problem in assuming that *Proconsul africanus* and *P. nyanzae* were simply males and females of a single species was that the sexual dimorphism of the teeth of the combined *Proconsul* sample was greater than the dental sexual dimorphism in *any* living species of Old World Monkey or ape. To my mind, the idea that *Proconsul* exceeded every other known primate in sexual dimorphism seemed highly unlikely.

The observation that a combined *Proconsul* species would have an unprecedented level of sexual dimorphism in its dental measurements convinced me and a large number of paleoanthropologists that there had to be at least two species in the genus *Proconsul.* Most were unwilling to accept that this fossil ape could violate the rules that governed sexual dimorphism in all higher primates. After considering this objection carefully, Jay went a step further in his hypothesis. Writing with David Pilbeam, who had already lumped *Proconsul* out of existence once, Jay proposed that exaggerated sexual dimorphism might be a biological characteristic of the genus *Proconsul.* In short, while they agreed that a combined *Proconsul* species had an excessive (by modern standards) degree of sexual dimorphism, they proposed that this variability was an unusual but genuine characteristic of the species.

This suggestion was anathema to me. If animals in the past played by biological "rules" that were different from those governing animals today, then all of us who study the past were in deep trouble. Without a belief in uniformitarianism, we were facing a blank wall when we looked at the past. Without uniformitarianism, we could no longer use the proportions, movements, habits, biology, or ecology of modern species as a lens for focusing on extinct species. If Jay and David were right, then no one would be able to verify or falsify any hypotheses, because there would be no known limits or biological "rules" against which to test the fossils.

To point up this problem directly, Mark Teaford, Chris Beard, Richard Leakey, and I published a paper on a fascinating specimen from Rusinga: the lower part of a *Proconsul* face, including ten teeth that Wambua had found in 1986. The incisors and canine teeth were missing from the palate, but the sockets of the teeth were well preserved and gave a good idea of tooth size. The large canine sockets suggested that the specimen was probably from a big male individual, yet the cheek teeth—which were still in the jaw—were intermediate in size between that expected for *Proconsul nyanzae* and *Proconsul africanus.*

So what species did the specimen belong in? We spelled out three alternatives. First, you could argue that it was a male *P. nyanzae,* with especially small cheek teeth and especially large canines. Though it is improbable that a male individual with unusually large canines would also have unusually small cheek teeth, it is not impossible.

Second, you could argue that this specimen was the first male maxilla of *P. africanus* to be found at Rusinga and was, coincidentally, a male with whopping great canines. Such might be the case, but then you were forced to wrestle with a most awkward fact, which is that *P. africanus* as it is formally defined on the basis of the Koru jaw does not show these dental proportions.

Although the Koru specimen has a large canine, it is not enormous like the one on the Rusinga fossil, and the large canine from Koru is not coupled with relatively small cheek teeth. Therefore, the new specimen from Rusinga cannot be a male *Proconsul africanus* because it is too different from the holotype in size and proportions to belong in the same species. The way we judge "too different" is based on comparison with living species. Any living species exhibits a characteristic range of variability in size, whether you are looking at tooth size, body size, anatomical proportions, or the size of some other body part. From such data, you can determine the range of variability in, say, molar size among living apes. Then if a questionable fossil specimen is so strongly different in size from the holotype that it falls outside the expected range of variability, you have to believe that the unknown specimen and the holotype are from different species. The second alternative identification of the palate, as a male *P. africanus,* can be ruled out on these grounds.

Third, you could argue that the specimen was a somewhat odd male *Proconsul nyanzae* of the type Jay envisioned, with all the specimens called *P. africanus* being the females of *P. nyanzae.* This argument is defensible if you are willing to accept that *Proconsul* showed more sexual dimorphism in its teeth than any living higher primate today—and I am not.

Funnily enough, there are three possible identifications for this specimen: two improbable ones and one impossible one. Why can't we agree on a species identification for a complete palate with most of its teeth in place? The most logical answer, I think, is that teeth are not very good indicators of species affinity in this instance. Teeth tell us the specimen is a male *Proconsul,* without debate, but they do not tell us which one.

Fortunately, there are many, many specimens of *Proconsul* from Rusinga that are not dental, so we have the unaccustomed luxury of consulting other parts of the skeleton from the limbs and torso as we try to sort out the number of apes in these fossil samples. What do the limb bones say about sexual dimorphism in *Proconsul?*

In skeletons of hominoid species that are alive today, the linear measurements such as length, breadth, or thickness of a bone in males is about 1.3 times the same measurement on the same bone in females. Expressed as a ratio, the sexual dimorphism of males to females on linear measurements is about 1.3:1. We compared these ratios to ratios constructed from measurements of various bones of putative *Proconsul nyanzae* and *Proconsul africanus* individuals. In the wrist bones, we found ratios that ranged from 1.5:1 up to a maximum of about 2:1, a degree of sexual dimorphism greater than that found in any single species of living hominoid. The wrist bones thus imply that two species are represented among the *Proconsul* remains. The same story was repeated when we measured the large bones of the ankle (the talus and calcaneus) of *Proconsul* from Rusinga and Mfangano.

Sometimes linear measurements are not as reliable an indicator as ones dealing with volume, such as body size. Technically, the attribute we commonly speak of as "body size" is properly known as "body mass," or the weight of an object under Earth's gravity. Body mass is the answer to the question *how big is it?* and it is a key piece of information about an animal. In fact, if I were offered my choice of any single bit of information about an extinct species, I might well select body mass as the most revealing. An animal's body mass scales with almost everything. By that I mean that if you know how big an animal is, you know a great deal more about that animal. Body mass varies in a predictable manner with a species' gestation length, its longevity, the spacing between subsequent births to one female, its home range, its tooth size, and its height. If you know the average body mass of a species, you can use a simple mathematical formula to make a good prediction of all of these parameters and more. Of course, body mass is not a straightforward concept because size often varies dramatically with sex. In living monkeys and apes, the ratio of male to female body mass may be as high as 2:1, as in baboons or gorillas. And sex, even in fossil apes, has a way of raising eyebrows.

I asked another colleague from Hopkins, Christopher Ruff, to work with me and others in estimating the body mass of *Proconsul* based on the nine femurs from seven different individuals that we had excavated from Rusinga and Mfangano. Chris is a meticulous scientist and obsessively careful about his measurements and analyses, which is an admirable trait. But when he is concentrating on a scientific problem, he becomes so focused that he can seem humorless, quite unlike his usual cheerful self. The plan was to begin by measuring living species of primates to establish the mathematical relationship between body mass and particular measurements. Because the femur supports the body, two measurements of the femur are especially indicative of body mass. One is the area of the cross-section of the femur at midshaft, which obviously must be strong enough to support the body as well as the force of muscle contraction. The other crucial measurement is the area of the articular surface of the head of the femur, which is the part of the thighbone that participates in the hip joint. Chris selected these particular measurements because they are so strongly related to body mass.

Whether the animal is standing or moving, its body weight passes down through the articular surface of the head of the femur to the leg and then to the feet. The body, as it is acted upon by gravity in any particular movement, generates a force that acts upon the articular surface of the femoral head. If the articular surface is larger, the force is spread over a larger area, meaning that the pressure is lower. Lower pressure means less wear and tear on the joint. The op-

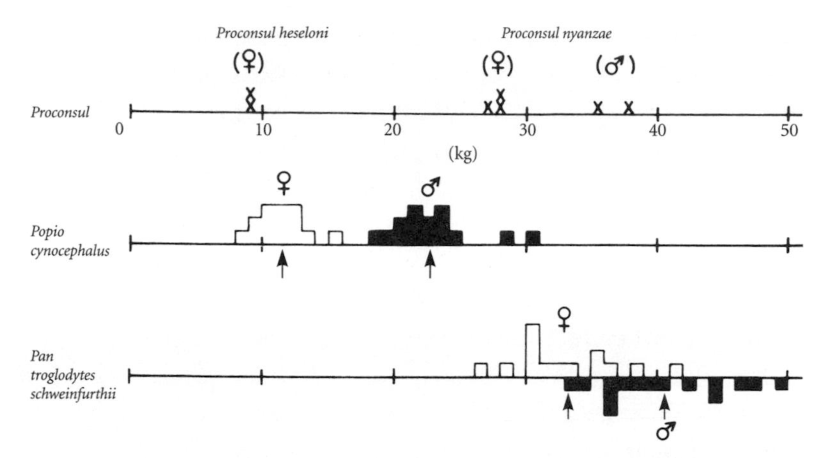

Our team estimated the body mass (in kilograms) of seven individual *Proconsuls* from measurements of their femurs. *Proconsul heseloni* had a body mass comparable to that of small female baboons (*Papio cynocephalus,* open boxes, middle histogram). *Proconsul nyanzae* females were comparable in size to either male baboons (black boxes, middle histogram) or female chimpanzees (*Pan troglodytes schweinfurthii,* open boxes, bottom histogram), while males of *P. nyanzae* were larger than baboons but comparable to chimps of either sex. The arrows below the middle and bottom histograms indicate female and male average body mass. (Modified from C. B. Ruff, A. Walker, and M. Teaford, "Body Mass, Sexual Dimorphism, and Femoral Proportions of *Proconsul* from Rusinga and Mfangano Islands, Kenya," *Journal of Human Evolution* 18 [1989]: 528.)

posite is true if the articular surface is smaller. A smaller articular surface means a higher pressure and potentially more damage to the joint.

In the initial phase of this study, what was really interesting was that the body masses of the living species fell into two distinct and separate clusters. The ape cluster ranged between about 55 and 85 pounds (26 and 38 kilograms) in body size and included both sexes of chimpanzees, gorillas, and orangutans. The monkey cluster centered on a body mass of about 20 pounds (9 kilograms) and included both sexes in two species of macaque. There was no overlap at all between these two clusters: apes were big, monkeys were smaller. It is well to remember, though, that there have been ape-sized monkeys in the past, though not at the place and time in which we were working.

Then we used these data on the femurs to estimate the body mass for the *Proconsul* specimens, using the equations we had generated in the initial phase of the study. The *Proconsul nyanzae* specimens yielded estimates ranging from 59 to 89.5 pounds (26.9 to 40.7 kilograms). These estimates fell into two clusters, one averaging about 59 pounds (26–28 kilograms) and the other about 80 pounds (35–38 kilograms). This pattern strongly suggests that these were females and males of a single species. The *Proconsul africanus* specimens yielded estimates ranging from 18 to 23 pounds (8.2 to 10.4 kilograms), with an average value of about 20 pounds (9 kilograms).

What was really curious was the distribution of the estimates compared to the data on the living animals. Particular fossil specimens consistently clustered with the monkeys, no matter which measurement of the femur was used as the starting point, and others grouped neatly with the apes. Those specimens that fell within the ape cluster had all been previously identified as *Proconsul nyanzae;* those that fell within the monkey cluster were all identified as *Proconsul africanus.* If all of these specimens belonged to a single species, then the ratio of male to female body mass was an alarming 3.4:1. Not only is this ratio much higher than the highest ratio recorded among living hominoids but it also exceeded the ratio of *any* living land mammal of any type.

"Such extreme variation would be difficult to accept within one anthropoid species," Chris as lead author observed soberly in the publication. For those who missed the obvious implication of that understated point, Chris added, "It therefore appears most likely that the smallest *Proconsul* on Rusinga and Mfangano (*P. 'africanus'*) is in fact a separate species from the larger group *(P. nyanzae).*" Chris put "*africanus*" in quotation marks because, by then, we had reevaluated the species *P. africanus*—a story I tell later in this chapter.

Kathy Rafferty, then a graduate student at Hopkins, decided to work with our team to enlarge the issue. Fossils were not going to be her specialty as a professional, but she hit upon a good question and followed it to an interesting conclusion. She wanted to determine the body mass of the biggest species of *Proconsul, P. major,* and find out how it compared with the body mass of *P. africanus* and *P. nyanzae.*

Proconsul major is the least well known species in the genus. Few of its teeth and even fewer of its postcranial bones have been found—and Carol had already shown that the few vertebrae attributed to *P. major* were unlikely to belong to a *Proconsul* at all. But there were still a few postcranial bones of *P. major* from the Miocene sites of Napak, Uganda, and Songhor, Kenya, that Kathy could use: a partial tibia and a couple of ankle bones. With help from the rest of us, she followed the same protocol and began by establishing the mathematical relationship between various measurements on these bones and the body mass of living species. Then she took the same measurements on the fossils, plugged them into the equations, and generated estimates of body mass. She compared these new estimates with those from our previous study of the talus and calcaneus in the two smaller species of *Proconsul.*

The dataset Kathy compiled made three distinct clusters of body masses: one including all of the purported *P. africanus* specimens, which ranged from 20.5 to 30 pounds (9.3 to 13.6 kilograms) and averaged 24 pounds (10.9 kilograms.); one for all of the purported *P. nyanzae* specimens, which ranged from 62 to 102 pounds (28.0 to 46.3 kilograms), with an average of 78 pounds (35.6 kilograms); and one for the purported *P. major* fossils, which ranged from 140 to 189 pounds (63.4 to 86.1 kilograms.), with an average of 165 pounds (75.1 kilograms). These clusters did not overlap and in fact there were large gaps between them, which suggests that they were meaningful clusters and not simply artifacts caused by random variations. Moreover, not a single specimen of the twenty-one fossils in the study had a body mass that fell into an unexpected cluster.

This project, once again, reinforced the contention that the postcranial bones of *Proconsul* made a strong argument for multiple species in the genus. Since previous studies of the teeth of *Proconsul* had consistently given more ambiguous separations among the species, Kathy's work was just one more piece of ammunition in the argument that the teeth of this genus were not very revealing of species affinities. Used to dealing mostly with teeth, many paleontologists found this conclusion unpalatable, but there was no way to deny the evidence from multiple projects. If you want to know which species a *Proconsul* belonged to, you shouldn't rely on its teeth to tell you.

The accumulated influence of all of these studies brought about a general consensus in the field that there were two different *Proconsul* species on Rusinga and Mfangano, in addition to the much larger-bodied *P. major* at the mainland sites. Even Jay Kelley came around to accepting that *P. africanus* and *P. nyanzae* were good species, although for a time he had been one of the most thoughtful and outspoken proponents of the idea that the two smaller *Proconsul* species from Rusinga and Mfangano were really only one.

While the field was struggling with these issues of sexual dimorphism and species identification, an odd thing happened. As various scientists examined the Miocene hominoid fossils anew and debated their interpretation, they started to suspect that there was an unrecognized difference *within* the specimens of *Proconsul africanus.* One after another, paleontologists began to see that the *P. africanus* material found on Koru and Songhor was different from the specimens discovered on Rusinga and Mfangano. These differences were subtle and based on details of dental anatomy because there were no skulls and only miserable scraps of postcranial remains of *Proconsul* from Koru or Songhor. There is no question that many of the small hominoid specimens from Koru, Songhor, Rusinga, and Mfangano all belong in the genus *Proconsul,* and yet there was something unsettling about them.

In 1993, four of us (Mark Teaford, Lawrence Martin, Peter Andrews, and I) undertook a virtual collaboration to fight our way through all of the evidence once again. Mark and I were in the same department, but the others were widely separated. Peter was at The Natural History Museum in London. He had written his Ph.D. thesis on *Proconsul* and other Miocene hominoids in East Africa many years before and had visited or worked at most of the Miocene fossil sites in Kenya; he has a particular interest in reconstructing the paleoecology of the animal communities at these sites. The fourth collaborator, Peter's former student Lawrence Martin, was at the Stony Brook University in New York. Lawrence's expertise was the structure of hominoid tooth enamel, and he had used dental features to classify and distinguish among the Miocene hominoids, not only of Africa but also of the rest of the Old World. Lawrence offered yet another perspective.

After much discussion, we wrote a paper announcing that, in our joint opinion, there were not three but four species of *Proconsul* in East Africa. We recognized the validity of *P. africanus*, which applied to the specimens from Koru and Songhor, *P. nyanzae*, and *P. major*. What was new was that we thought the small *Proconsul* from Rusinga and Mfangano, formerly considered to be *P. africanus*, was a new species that included the magnificent *Proconsul*-in-a-tree as well as Mary's beautiful little skull. In that paper, we pointed out the anatomical distinctions between the Koru-Songhor *P. africanus* and the specimens of the small *Proconsul* from Rusinga and Mfangano. We named the new species *Proconsul heseloni*. I was deeply satisfied with the naming of this new species on two counts.

First, it provided an opportunity to honor Bwana Heselon Mukiri, one of Louis Leakey's great Kikuyu fossil-finders and the man who helped Mary Leakey excavate the 1948 skull from Rusinga. In his correspondence with Le Gros, Louis always called the little skull "*heseloni*" and I felt it was right that this informal designation should become the formal one. Heselon Mukiri was a man of talent and knowledge whom I was delighted to honor.

Second, we had the chance to designate a new holotype for the new species. Because it is the most complete early Miocene hominoid yet known, possessing the greatest number of different body parts to facilitate comparison, we made the *Proconsul*-in-a-tree skeleton (KNM-RU 2036) the holotype of *Proconsul heseloni*. In the hypodigm—the formal statement of specimens defined as belonging to the new species at the time of naming—we included Mary's little skull and the huge number of specimens from Rusinga and Mfangano, including the articulated hands and feet from the Kaswanga Primate Site. This was an impressive holotype and an abundant hypodigm that should enable a future scientist to compare a new find to the equivalent body part in *P. heseloni.*

This is not to disparage what Hopwood had done 60 years earlier. As is often the case, Hopwood had very few fossils to work with when he named *Proconsul africanus* in 1933. He picked the best specimen he had, the partial male jaw from Koru, and made it the holotype of *P. africanus.* It was certainly not Hopwood's fault that this holotype posed real difficulties when other parts of the skeleton began to turn up. How can you determine if a new fossil is the same species as a previously named one, if the two have no parts in common? The answer is that you can't; you can only guess, and we had all done so for years.

You may think the questions we raised and the analyses we went through to get *Proconsul* sorted out constituted an idiosyncratic problem, but its twin is found over and over again in paleontological studies of every age and in every part of the world. What defines a species? How much variation can we expect within a species? How can we tell males from females, using only the fragmentary fossil remains that we have? What is the best thing to do with distinctive but incomplete remains? These are classic problems we encounter when trying to interpret the past.

There are several keys to answering these questions, but they are not simple ones. Uniformitarianism is a powerful tool and the present can certainly be used to understand the past. But, as was shown in the lengthy and serious discussions over the sexes and species of *Proconsul* and how they are to be distinguished from one another, the correct way in which to use the present to interpret the past is rarely obvious. I prefer to believe that the presently living animals, plus those that have sadly become extinct in the very recent past, exemplify much about the way in which species vary in terms of fundamental biological and mechanical principles. For example, I have no qualms about saying that a knee with a particular shape to its joint indicates that a particular type of locomotion—such as rapid jumping—is both habitual and important to the species that owns that knee. I believe firmly that certain overarching principles of biology applied as strongly in the past as they do in the present. These are principles such as: sexual dimorphism in mammals can be carried so far and no farther; or, the length of the period from birth to sexual maturity in a species is closely correlated with its body size. Yet I am willing to accept that in other ways the past is different from the present. I am intrigued but not dismayed that all the evidence available points to the fact that apes were ridiculously abundant and diverse in the Miocene of Africa while monkeys were rare, even though this is the opposite of the situation today.

My aim, which I learned from John Napier when I was a fledgling graduate student, is to bring the past back to life, to envision and understand the biology of the past as well as I can that of the present. I will use every means at my disposal to restore a colorful, detailed, and rich image of what the world was once like. I am greedy in this, for I want to know everything, every tiny fact that can be squeezed out of the available evidence. At each step, though, due thought and care must be applied. Even standard and well-established methods of analysis cannot be applied thoughtlessly. New methods—new analyses, new comparisons, and new techniques—are invented every year to solve particular problems. They bring with them both new insights and new pitfalls.

Those of us who are obsessed with bringing the past back to life walk a tightrope every day. On the one hand, we do not wish to blunder and underestimate the wonderful uniqueness of the past. There is nothing I enjoy so much as a secure but surprising result. On the other, we would err as badly if we overemphasized or overestimated the differences between the past and the present. The process of finding the right path, of staying on the tightrope, sometimes confuses onlookers because researchers often do not agree where that slender thread of truth and insight lie. Sometimes people who are too used to taking an authority's word for things are dismayed that The Experts can't agree. The very process of debate and disagreement—which can be likened to winds that can blow you back onto or off of the tightrope—fosters careful thought. In the end, though, it is the difficulty of trying hard to discern the truth that leads to good science and keeps us honest and on course.

9

How Many Apes?

If I thought for a moment that I understood the role of the Miocene apes in our family tree simply because my collaborators and I had sorted out the number of *Proconsul* species and their adaptations, I was sadly mistaken. While we were working on the specimens from Rusinga and Mfangano, new hominoid fossils were being found at a dizzying rate at other sites in Africa, Europe, the Middle East, and Asia. The litany of names for Miocene hominoids and of the sites from which they come is bewildering, so I will only mention a few.

I was involved in finding one of the strangest Miocene hominoids at a site in northern Kenya known as Buluk, which is northeast of Lake Turkana just south of the Ethiopian border. Richard and Meave Leakey went there for a quick survey in 1983 because a geology student who was mapping the area had reported that there were fossils at Buluk. When they too found fossils, Richard asked that I drive in with a few of the Hominid Gang from Koobi Fora, the permanent camp Richard had created on the east side of Lake Turkana as a base for expeditions. This was a year before we started work on Rusinga.

The plan was that three of us—Maundu, Wambua, and me—would set out from the east side of Lake Turkana to drive about 50 miles cross country to Buluk. Once there, I returned to Koobi Fora for more men and more camping and excavation gear, while the others marked out an airstrip on reasonably level ground so Richard and Kamoya could fly in with more supplies.

The very first day our small expedition arrived at Buluk three of the local tribesmen appeared. They were tall and very dark skinned and, to put it bluntly, almost stark naked. Their outfits consisted of World War II Enfield rifles and bandoliers full of ammunition, hide sandals, peculiar decorated belts draped about their waists, and a length of cloth that one of them had, which he usually kept wound turban-style around his head. When I looked at the belts closely, I realized they were leather strings from which dangled peculiar wrinkled objects: the testicles of the men they had killed. They watched us intently and openly. Their stares and their belts made us rather nervous.

In hopes of establishing some sort of friendly relations, we tried to speak to them in each of the languages we knew collectively: Kiswahili, Kikuyu, Kikamba, English, French, and German. I even tried Italian, since Ethiopia had been an Italian colony briefly during the twentieth century. "Parliamo Italiano?" I asked hopefully, thinking that if they did, I'd try to speak to them in my schoolboy Latin. They squatted and simply looked at me with blank eyes.

We decided to get on with setting up the camp while unobtrusively keeping an eye on them. They watched us, stone-faced but I think fascinated, as we erected houses out of canvas—tents to live in, carefully spaced to take advantage of what shade there was and to provide some privacy. As far as we knew, the local tribe to which they belonged probably built inverted bird's-nest-style huts of grass and other vegetation, like the Dassenech or the Turkana people who lived nearer to Lake Turkana. The largest tent, shaded with a tarpaulin and with a canvas floor, was the mess tent, where we would eat and work on fossils. (You always want a clean, smooth floor in the place where you work on fossils, in case you drop a fragment and need to find it again.) We leveled the large dining table and placed it with as much of a view as we could find; these little touches somehow make a great deal of difference in the field.

When they saw us unloading drums of water, they gestured to the large, empty gourd they were carrying and mimed drinking. We let them drink their fill and replenish their gourd, and then they walked off into the desert. All of us were rather nervous while these strangers were in camp because their ways and manners were so utterly remote from the ones that were familiar to us. We didn't see them for some time after that encounter.

I drove back to Koobi Fora to pick up more men and supplies. When I returned to Buluk with Solomon the cook and his brother Aila sons of Derekitch, they were keenly interested in our visitors. Solomon and Aila were Dassenech, members of the tribe that lived on the eastern shore of Lake Turkana, and they reckoned they could work out what tribe the three men were from. I described the way the trio styled their hair and the belts of trophies around their waists. Aila and Solomon began talking rapidly to each other in Dassenech and gesticulating. They recognized the strangers as members of a tribe they called the El Malakok, more commonly known as the Hamar, or Hamar-koke. The Malakok were deadly enemies of the Dassenech and hence of Aila and Solomon.

"Very bad men," Solomon told me seriously, looking grim. *"Very bad."*

I admitted that they looked like very bad men to me too. The fact that we could communicate with them at only the crudest level worried me a lot. Since anyone could remember, the Malakok had engaged in an unceasing series of bloody raids and counter-raids with the Dassenech and Turkana peoples over women, cattle, and goats. The warriors ranged back and forth across the border between Ethiopia and Kenya, using the unmarked and wholly insignificant (in their eyes) political boundary to protect them from the police or military forces that might try to capture and punish the raiders. They were not known for showing any mercy to their enemies—witness the belts.

A day or so later, Richard and Kamoya flew in with more supplies and men, but Richard was needed back in Nairobi and had to leave almost immediately. Aila and Solomon seemed jumpy and watchful and made a point of telling Kamoya about the Malakok. We thought we'd seen the last of them. The gang and I had been talking for days, trying to decide if their coming into camp was a preliminary to slaughtering us in our beds as we slept. Aila and Solomon generally thought the worst; the rest of us were uncertain and uneasy. Their identification as the most bitter enemies of the Dassenech, when we had two Dassenech in camp, was no comfort.

I had found a beautiful gomphothere skull—a gomphothere is another sort of extinct elephant—with just the top of the crest along the top of its skull, where powerful jaw muscles once attached, showing above the sediments. Kamoya and I went out the next day to excavate it while the others prospected for more fossils. To get the skull out without damage, we began by excavating a large cavity all around the fossil, which we were lying in quietly as we worked together in sociable silence the way you can with an old friend. We could hear goats bleating somewhere not far away, a lot of them, the tinkle of wooden cattle bells, and the faint cries of sheep. The air was full of dust but I didn't think anything of it. Then I caught a movement of something dark-colored out of the corner of my eye.

"Don't get up," I said quietly to Kamoya, "and don't make any sudden moves. We are not alone." I moved my eyes in the direction I wanted him to look and he saw, as I had, that one of the Malakok was standing on the edge of the gully just above us. He had a rifle in his hand and was watching us.

"This is the man who worries Aila and Solomon?" Kamoya asked softly. I nodded, my heart beating faster and my sweat flowing more freely. We could do nothing, so we pretended to ignore him. After a while, he left and we breathed easier.

We were really happy to get back to camp for lunch that day, but when we started telling the others about what we had seen—the Malakok and all the herds of goats—Aila and Solomon grew frantic.

The camp at Buluk, where in 1983 we encountered the three Malakok brothers, members of a tribe renowned for ferocity. That season, we found the first specimens of *Afropithecus*. (© Alan Walker.)

After a hurried exchange in Dassenech, which no one else understood, Aila and Solomon took off at a dead run. The rest of us called after them but they did not stop to answer. We were shaken and very puzzled.

The explanation came hours later. Goats, to the Dassenech people, are the most desirable and precious commodity in existence. Aila and Solomon knew from the dust and the sounds that someone was moving a very large herd of thousands of goats from the southwest, where their people lived, to the northeast. Fearing the worst—that their own goats had been stolen and their relatives killed—the Dassenech brothers had dropped what they were doing and run off to check on the goats. They ran until they reached a place where they could sit, silent and suspicious, to watch the herds that were being moved, prepared to battle to the death if any were their own. Aila and Solomon scanned the herds of goats pointedly, satisfied only when they knew for certain that none of them were theirs. They also saw many women and children and lots of men none of us had ever glimpsed, who had stayed out of sight while the Malakok kept us under surveillance.

After a while, the three Malakok spotted the Dassenech and came over to speak to Aila and Solomon. Using their few words in common and gestures, Solomon bartered his shirt for a fat goat, which they brought back to camp for dinner as a sort of peace offering.

Kamoya and I bawled Aila and Solomon out for their potentially dangerous actions. They had taken off precipitously without telling anyone and, worse than that, they had risked drawing us all into a fatal conflict. It was only sheer luck that none of the goats in the Malakok herd had been stolen from their families.

They looked suitably chastened but I knew if they had seen goats they felt were theirs, we might all have been killed. The value of goats was too deeply ingrained in them to be eradicated by a mere reprimand. After receiving their promise that they would not be so foolish another time, I had to admit that we were glad of the fresh meat. I even suggested they conduct a similar barter every few days, since we had plenty of shirts.

After three shirts had been traded for three goats, the market was saturated. The Malakok—we thought they were probably brothers—had no need of additional shirts and would not consider trading a valuable goat to enlarge their wardrobes past their immediate needs.

From then on, the Malakok trio turned up almost every day to watch us. We would think they were gone and then we'd catch just a tiny movement and we'd spot a figure or two lounging silently in the inadequate shade of a thorn bush and watching us. It was incredibly menacing.

One day just before lunchtime the three men moved up more boldly and watched fixedly as Solomon started a fire and set to work putting together lunch. I was in camp a bit early and kept a close eye on them and on Solomon, who I knew was very nervous about the Malakok. Solomon had baked fresh bread in the fire, using an old ammunition tin as an oven—Solomon's bread is wonderful—and he set it out on the table. Then he boiled water for tea and soup and opened tins of beans and corned beef to warm.

The three Malakok stared in open astonishment when Solomon opened the tin cans and scooped out food. That alone told me that these three men were wilder and less in touch with civilization than anyone we had ever encountered before. Of all the remote places the Hominid Gang and I had been in East Africa, we had never before met men who didn't know what a tin can was.

I decided to try to solidify friendly relations. I gestured to the Malakok to come into the mess tent, hoping that this was not going to prove a fatal error. They stopped, unsure, at the edge of the canvas floor, then carefully took off their sandals and stepped onto it reverently as if it were a fine Persian carpet. The youngest bent down to inspect its smooth surface. One of the men took over three camp chairs for them to sit in and they sat down hesitantly in the chairs, which fortunately did not collapse on them as camp chairs sometimes do. They stroked the planed wood of the arms of the chairs and fingered the canvas. One of them nodded to us in approval of the fine quality of our goods, when in fact they were nothing but the cheapest canvas folding chairs we could find in Nairobi.

I knew I was probably the first white man they had ever seen. More shocking than that, these men had apparently never seen tin cans, canvas, tents, chairs, and perhaps did not know about enameled metal mugs and plates. This was a situation that could go well or very badly indeed.

I asked Solomon to give them big mugs of orange-colored tea, sweetened with lots of sugar and condensed milk the way most Africans take it, and to offer them food. Solomon was reluctant to be hospitable to his enemies but complied, a little sour-faced, when my expression told him I was serious. The Malakok took the enamel mugs, sniffed the liquid, and watched us drink the same substance. Perhaps they were familiar with tea, because soon they decided it was not poisoned and gulped it down and asked for more. When the plates of food came, with spoons, they stared at us again in indecision. The stuff on their plates smelled like food. We were eating what appeared to be the same thing. But was it safe to eat? The three men talked and gestured among themselves, and then, under pressure, the youngest carefully, suspiciously, took a small bite. Fortunately he found nothing amiss; in fact, after the second bite, he began scooping beans and meat into his eager mouth with enthusiasm. The others followed suit and, after eating and drinking so much that their normally lean bellies seemed swollen, they settled back in their chairs and stretched their legs out in front of them with contented expressions on their faces.

"Kamoya," I said, sotto voce. "Have you seen the feet on that one?" I gestured very slightly with my head.

"Yes, Alan," Kamoya replied quietly. "I have never seen a man with such large feet."

We tried to keep straight faces, for it would not promote good feelings if we laughed at these men and insulted them. The size of the one man's feet was truly amazing. They were flat and wide, with little arch, as the feet of people who have never worn shoes often are. They were also *huge.* We took to calling him Big Foot.

Our visitors soon loped off across the dry landscape, striding confidently to who knows where. They continued to come back and watch us every single day and we continued to be wary. Sometimes I looked out into the absolute darkness that surrounded us at night and felt them there. I knew it was not possible that they were alone; all of those people we had seen that one day had to be somewhere close by. Some days we would think our activities had lost their appeal to them and Big Foot and his brothers hadn't bothered to come. Then someone would spot them half-concealed behind a rock or bush: watching, watching, watching.

Watched or not, we prospected for fossils every day, but no one wanted to get too isolated from the others in case Big Foot and his brothers decided to attack. They gawked as we put plaster casts on fossils, using a strange white powder that turned to stone: something they seemed to find miraculous. What they made of our activities, I will never know. Even though Aila and Solomon eventually managed to find a few words that they had in common with these men, asking what they thought of fossils was far beyond anyone's linguistic or mental abilities. Most of the time they seemed as impossible to communicate with as a tree or the vultures that flew overheard. We had very little in common.

A few weeks later they saw Richard fly in with supplies again. They must have seen airplanes flying high in the sky, but they could never before have seen one close up. We did not see the Malakok that day but I knew they were somewhere observing the airplane and us. Richard was very impressed with our stories of Big Foot's pedestrian anatomy. I decided to try to get a plaster cast of his foot somehow. It was a silly lark, perhaps my reaction to the tension we had been living under. All I could do was to conjure up a special ceremony from our tribe—the one that Richard and I and the Hominid Gang constituted—and ask Aila to try to convey the concept to Big Foot and see if he would cooperate.

He was to tell them that we were leaving the area soon and would leave them gifts as a symbol of our friendship and appreciation for their not killing us. We would give them a good knife, some blankets, and all of the tins and bottles that we had emptied, items which to a people without artificial containers of any sort are highly valuable. And, I asked Aila to explain, we wanted to conduct a special ceremony to seal our friendship for all time. In our tribe, two chiefs who trusted each other went through a special ritual. They would each sit in the very fine chairs we had and drink a special drink. Then, each man would place his foot, side by side with the other man's foot, into a large container that held a few inches of the magic powder that turned to stone. The powder would grow warm with the warmth of their friendship and would then turn to stone, as they had seen it do before. Then the two chiefs would remove their feet and the hardened plaster plaque. The plaque would be broken down the middle, each chief giving to the other the cast of his own foot to keep. Should they meet again, those casts would match the other's foot and show they were allies. Our chief was Richard, the man who came in the airplane. He wanted to exchange footprints with their leader, Big Foot.

The gang were very dubious that anyone would believe this ridiculous fairy tale and I had to admit it wasn't very convincing. Aila conveyed this story to Big Foot as best he could and asked him to come back on the following day, when our chief would fly in for the ceremony and the gift giving. Neither Big Foot nor his brothers showed up. We left the tins, the knife, and the blanket in a prominent place and hoped they would find them.

I thought the fictional tribal ritual was the last I would ever hear of Big Foot, but I was wrong again. One night about a year later, a runner came panting into Koobi Fora looking for help. It was the middle of the night and Richard was in camp, sound asleep, but the exhausted man insisted on seeing him immediately. He had come as fast as he could from Buluk, some 50 miles away, where there was a deadly battle in progress. As usual, Big Foot and his tribe had continued to raid villages on the Kenya side, retreating back into Ethiopia to escape retribution. But a few nights earlier, they had made the terrible mistake of attacking and wounding some of the Kenya Police, who were now exacting their revenge in an intense firefight. Big Foot had sent a runner to tell Richard to call off "his" men, saying that he would surrender and accept his punishment if only the police would stop the killing.

There was nothing Richard could do. The Kenya Police were not under his control and he did not have any way of calling them off. Besides, if the fighting had been bad when the runner left Buluk, probably not one of Big Foot's men was still alive by the time the runner reached Koobi Fora. We had been perhaps the first real contact with the wider world that Big Foot and his brothers had had; now they were dead.

If the trip to Buluk was unsettling, the primate fossils we found were exciting. They were hominoids, fairly large-bodied ones, and we had a well-preserved lower jaw and part of an upper jaw, with many teeth in place, and a few isolated teeth: good specimens. Back in Nairobi, Richard and I carefully compared our new material to other known hominoids. The closest thing we could find to our new fossils was a genus from Pakistan and India known as *Sivapithecus*, which may not be the direct ancestor of orangutans but is certainly related to them. All the teeth on our fossils were very thick-enameled and the canine teeth were especially stout and fat; the jaw itself was long and rather narrow. These were all *Sivapithecus* traits. But the only unequivocal specimens of *Sivapithecus* were found on another continent and were only 10–8 million years old, so this new Buluk creature, which was about 17 million years old, was at the very least a new and older species in the same genus. But despite the quality of the specimens, we were quite mistaken in thinking the Buluk hominoid was a new *Sivapithecus*.

It wasn't until Richard and Meave worked at Kalodirr, a site of the same age as Buluk but on the west side of Lake Turkana, that we realized we had misunderstood the Buluk specimens. Kalodirr yielded hominoid and other fossils that showed us how easily even experienced paleontologists can be fooled by fragmentary specimens. Wambua found two spectacular partial skulls, consisting of most of the faces of a male and female hominoid. Others found more teeth and jaws that closely resembled the Buluk specimens, as well as some postcranials. Although our preliminary assessment of the Buluk hominoid had been based on perfectly good anatomical evidence and what seemed to be good specimens, too, we were wrong. Many of our colleagues agreed with us that we had found an important link between the early Miocene African hominoids and the later Miocene Asian ones. The additional finds proved that, although the teeth and jaws that we found at Buluk looked like *Sivapithecus*, that and general body size were the only points the two had in common.

Richard and Meave named the new ape *Afropithecus turkanensis*. The *Afropithecus* skull has an extraordinary long, narrow snout and a face that, in profile, slopes outward all the way from its brow ridges to its procumbent front teeth. To be sure, the profile of Mary's little

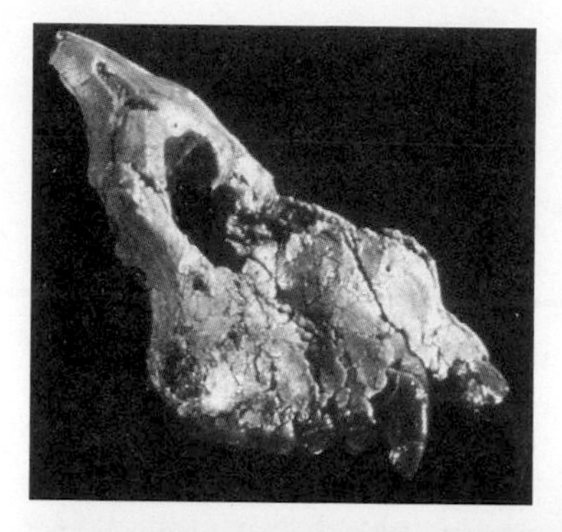

The facial skeleton of *Afropithecus* (bottom, lateral view), a 17-million-year-old species, closely resembles that of its possible ancestor *Aegyptopithecus* (top, lateral view), a 33-million-year-old species. Images not to the same scale. (*Afropithecus* photo by Richard Leakey, © National Museums of Kenya; *Aegyptopithecus* photo courtesy of and © Eric Delson.)

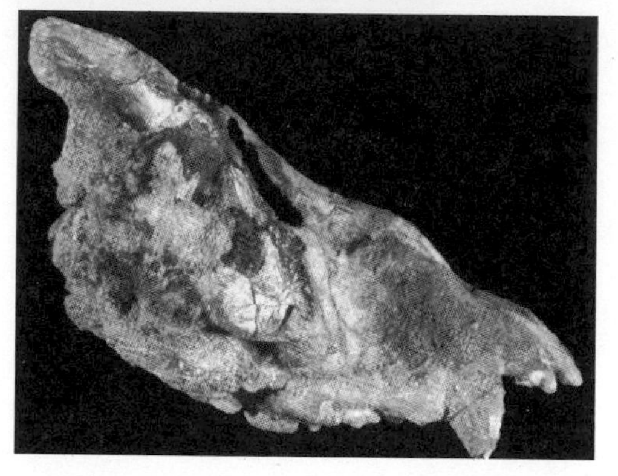

Proconsul skull slopes outward from top to bottom, but compared to *Afropithecus, Proconsul* has a fairly short face that is tucked much more underneath its braincase. It is as if a metaphorical and overwrought designer thought of the principle of a long sloping face and then carried it to an extreme in *Afropithecus.* Little of its cranial vault is preserved, but the palate is intact and contains all its teeth, including stout and stubby canine teeth, which are found in both males and females of this species. The best skull is from a male individual, but there is also a partial female cranium and a pair of upper and lower jaws from yet another individual.

Afropithecus was clearly a large-bodied ape, perhaps the size of *Proconsul major,* and the two have similar-sized teeth. However,

Afropithecus has a much narrower snout and a much higher face than *P. major*, which suggests that the overall size of the skull of *Afropithecus* would have been larger than that of *P. major*. We don't have a complete braincase, but the general look of the skull suggests that the brain was small. The postcranial fossils of *Afropithecus* are similar in proportion to those of *Proconsul nyanzae* and resemble them in some functional features, which indicates that the new species may also have been a slow-climbing arboreal quadruped.

More than anything else, the *Afropithecus* skull resembles the skull of a fossil primate called *Aeyptopithecus zeuxis* from the Fayum region of Egypt. Why isn't the skull *Aegytopithecus* then? One problem lies in its size. *Aegyptopithecus* was a much, much smaller animal than *Afropithecus*. The most complete *Aegyptopithecus* skull is tiny, just over 3 inches (about 7.5 centimeters) long from the tip of its snout to the back of its head. Without most of its braincase, the face of *Afropithecus* is almost 5 inches (about 13 centimeters) long. With a braincase intact, the head of *Afropithecus* must have been another inch or two longer—and of course much larger in volume—than that of *Aegyptopithecus*. The second difficulty is the fact that *Aegyptopithecus* is from the early Oligocene, a little more than 33 million years ago, which makes *Aegyptopithecus* nearly twice as old as *Afropithecus*. For a single species to persist for 16 or 17 million years would be remarkable. Yet the resemblances between the two are striking enough that we couldn't help speculating that *Aegyptopithecus* may have been ancestral to *Afropithecus*. Of course, it is possible that this type of face was simply primitive for all early monkeys and apes. Only more fossils will let us choose between those alternatives.

When I try to reconstruct the habits and adaptations of *Afropithecus*, my best hypothesis is that it was probably a seed predator. A living primate species that fills this ecological niche is the saki monkey of South America, for example. These monkeys have strong, stout jaws, broad canines that vary little between males and females, thick-enameled molars for cracking open tough nuts and husks, and stout, strong, procumbent incisors for peeling the coverings from seeds. Each of these anatomical adaptations to seed predation is found in *Afropithecus,* though its estimated body weight is much heavier than that of the living examples among primates. However, absolute body weight is no barrier to being a seed predator. Among nonprimates, tapirs are well known to be seed predators and they weigh 400–600 pounds (200–300 kilograms)—much more than *Afropithecus* could possibly have weighed.

Because *Afropithecus* is an arboreal species that is relatively unspecialized in its limb bones, and also has very thick enamel—both features found in many of the middle Miocene apes from Europe—*Afropithecus* is a reasonable candidate for being one of the few Miocene hominoids to disperse out of Africa and reach Eurasia about 16.5 million years ago.

Kalodirr also yielded yet another Miocene ape: *Turkanapithecus kalakolensis.* Like *Afropithecus, Turkanapithecus* is a hominoid represented by a good number of specimens, but the prize is a partial skull. Kalodirr has been radiometrically dated to between 16 and 18 million years ago, which makes these apes about the same age as the *Proconsul*s from Rusinga and Mfangano and *Afropithecus,* too.

The skull of *Turkanapithecus* is probably male, because it has large canine teeth. *Turkanapithecus* has a distinctive muzzle that is squared at the front with a broad opening for the nose; its bony orbits are widely set apart, and it has clear browridges over the eyes. In all these features the skull of *Turkanapithecus* differs from that of *Proconsul* or *Afropithecus.* Like the *Proconsul* specimens, though, *Turkanapithecus* shows an intriguing mixture of apelike and monkeylike anatomy in its limbs and body.

The third primate species discovered at Kalodirr was *Simiolus enjiessi.* The second part of its name may sound like a word in some African language, but it is in fact a small joke: a phoenetic spelling of NGS, the acronym of the National Geographic Society, which has funded many of our expeditions.

Simiolus differed strikingly from *Afropithecus* in the important re-

This enigmatic little skull of *Turkanapithecus* (anterior view) has a squared-off muzzle, distinct brow ridges, and a wide space between the orbits, unlike *Proconsul* or *Afropithecus,* yet all three species are 17–18 million years old. (Photo by Richard Leakey; © National Museums of Kenya.)

spect that it wasn't an ape or a hominoid at all. To understand what *Simiolus* was, you need to think about how species are classified. The procedures of taxonomy (the practice of classification) are rigorous and carefully spelled out, though their application can be controversial. The ideal classification reflects the reality of evolutionary groupings and can be replicated by any careful individual to produce the same classification of specimens. It is fair to say that those lofty aims have not yet been achieved.

The most widely used system for classification today is cladistics. According to this method, the classifier's first step is to make note of as many observable features or traits of the specimen as possible, such as the numbers and types of teeth and the location of different anatomical features of the skull and limb bones. The traits listed may number into the hundreds if the specimens are largely complete, or may be a handful if the specimens are fragmentary. Obviously, the resultant classification is more likely to be reliable if a larger number of traits is used. Ideally, each trait is a "yes-no" feature: one that is either present or absent rather than being present in varying degrees among individuals in a population. Unfortunately, most higher primates in the Old World have basically the same skeletal plan—the same numbers and types of teeth, digits, and limbs, and the same bones of the skull—which means that there are few "yes-no" traits.

To determine the relationships among a whole group of specimens, these lists of traits for each individual are put into a computer. A specialized program is used to cluster each specimen with the one that it most closely resembles; the cluster of two is then grouped again, with the specimen or group of specimens with which it shares the most traits, and so on. The result is a nested set of categories that are displayed as a branching diagram known as a cladogram. At each branching point is a node, or point of divergence, characterized by a set of features that arose evolutionarily at that point. Cladograms are usually read left to right, so that all specimens listed to the right of a node possess all the features that characterize that particular node plus all the features characterizing every node to the left of the specimen's point on the cladogram. The more primitive, ancestral species thus fall to the left of the cladogram and the more derived or specialized ones fall to the right. It is very tempting to take a cladogram as a diagram of evolution, but it really only shows the general pattern of branching that has occurred among the characters, *not* the branching pattern among the species. A cladogram must be redone any time any specimen is discovered that reveals a new feature or set of features. Theoretically, at least, at some point a cladogram and a phylogeny—a diagram of evolutionary relationships—converge and become identical, but with fossils theory can be a long way from truth.

The drawbacks of cladistics are three. First, the result is overwhelmingly influenced by the selection of traits that are enumerated. I'll give an example. Suppose species X has very big, thick-enameled molar teeth. Is this a single trait (large teeth) or four traits (long teeth, wide teeth, tall teeth, thick-enameled teeth)? The answer depends upon whether or not the person classifying the specimen thinks that the enamel thickness, length, height, and width of teeth are independent variables. Does a tooth that gets longer because of a genetic change also inevitably get wider? In that case, length and width are not two independent traits but represent a single genetic change. Or are length and width uncoupled genetically, which would mean that they really are two separate traits? Usually we do not know if traits are independent and must guess. Then there is the question of how "longer" or "wider" or "thicker" is to be defined. As the old computer saying goes, "Garbage in, garbage out." The cladogram is only as good as the trait list. Second, cladistics recognizes only evolutionary change that causes a splitting or speciation event. Thus if a single lineage remained fundamentally the same, interbreeding population through time, but evolved smaller size, this change would not be recognized in a cladogram until the size change was so dramatic that a late specimen was considered to be a different entity (species) from an early one. This mode of evolution is known as anagenesis and simply cannot be captured by cladistic techniques. Third, cladistics usually takes no account of time, so an important variable in assessing any fossil must be ignored.

Problems aside, cladistics is the best approach anyone has devised yet, and so paleontologists will use it until something better comes along. It is to cladistics we have to turn to try to answer the questions: What is an ape? What is a monkey? Was there ever a creature closely related to both apes and monkeys that was neither one?

All apes and monkeys from the Old World (Africa, Europe, and Asia) fall into the large taxonomic group known as the Catarrhini. This group also includes humans, because the humans and their ancestors share key anatomical traits with the Old World monkeys and apes. Even though many humans now live in the New World, the unmistakable imprint of that anatomical heritage tells us that the human lineage arose in the Old World. So do the fossils, but that is another story.

If we ignore humans, the living catarrhines can be subdivided into two main groups: (1) monkeys and (2) apes, which are also called hominoids. But before catarrhine monkeys became monkeys and apes became apes, there were simply basic, primitive Old World higher primates that are sometimes called "stem" catarrhines. The stem catarrhines were the stock from which both monkeys and apes later evolved.

What evolutionarily novel traits or anatomical features distinguish Old World monkeys from apes? What makes an ape an ape?

We can make a partial list of the more obvious features that would show up on Old World fossils. Monkeys have tails; monkeys have sitting pads and the underlying skeletal structure to support them; monkeys do not have sinuses in either their frontal bone (the bone over the eyes) or the maxillae (the bones underlying the cheeks); monkeys have more lumbar vertebrae than apes; monkeys have more flexible backs than apes owing to the shape and placement of the processes on their lumbar vertebrae; monkeys have teeth and skulls that differ in tiny anatomical details from those of apes; monkeys are generally quadrupedal in their locomotion, which involves many adaptations in their limbs and digits; monkeys have deep, narrow chests and pelvises. Apes do not have tails; apes do not have well-developed sitting pads (though lesser apes, the gibbons and siamangs, have small ischial callosities and are thus judged more monkeylike than other apes); great apes have frontal and maxillary sinuses, though lesser apes have only maxillary sinuses; apes have fewer lumbar vertebrae than monkeys (with gibbons and siamangs being, again, intermediate); apes have stiffer backs than monkeys owing to the shape and placement of the processes on their lumbar vertebrae; apes have teeth and skulls that differ in anatomical details from those of monkeys; apes are either habitually involved in suspensory locomotion in the trees or their skeletons reveal a recent ancestry of such locomotion; apes have broad, shallow chests and pelvises.

In addition to all of these traits in which there are differences between apes and Old World monkeys, there are a huge number of traits that are held in common by both groups. These are the primitive traits that make up the bulk of all observable features on any being. For example, all primates and a great many other mammals as well as reptiles have a head, a trunk, and four limbs with five digits on each; all female mammals have mammary glands to feed their offspring; all mammals have (or did have) hair or a very hairlike body covering. These traits are primitive—so primitive that they are not useful in sorting out the fine points of primate phylogeny. Derived or specialized traits are the ones that may demarcate the ways in which a species or evolutionary lineage has diverged from its close relatives. Thus, for example, both gibbons and siamangs have greatly elongated arms; elongated, curved, and hooklike fingers; and extremely mobile shoulder joints. These are but a few of the derived traits that have evolved in gibbons and siamangs as they have become specialists in brachiation. Because these traits are common to gibbons and siamangs, they suggest that there was a stem lineage that gave rise to both and that these adaptations occurred after the stem lineage diverged from the lineages of other apes and before gibbons and siamangs diverged from each other.

However, we have to consider the possibility that what accounts for the anatomical resemblances between gibbons and siamangs is not close relatedness but convergence. Possibly (but in this case, most improbably) gibbons and siamangs evolved these adaptations in parallel—independently, in two separate evolutionary events—because they shared a common lifestyle. Considering the details of exactly how one species is adapted for, say, brachiation through the trees as compared to another species may reveal whether the two have taken a common evolutionary path. To continue with the same example, South American spider monkeys also have very long arms, hooklike hands, and extremely mobile shoulder joints because they, too, swing through the trees. Are spider monkeys also closely related to gibbons and siamangs? The overwhelming consensus is that the spider monkey did not evolve its suspensory adaptations in common with gibbons and siamangs because there are so many differences between Old World apes and New World monkeys. For example, the New World spider monkey has a long, prehensile tail that it uses for suspension while the Old World gibbons and siamangs are tailless;

spider monkeys have a different number of teeth from gibbons or siamangs; and, tellingly, the Asian apes differ from the South American monkey in the details of how each has adapted its limbs and joints for suspensory locomotion. This, then, is a case of convergence. But we have very few secure instances of close convergences, which means that we have no good idea of how rare or common they are. Convergence is the nightmare of cladists, the unexpected occurrence that could instantaneously transform a perfectly good cladogram into nonsense.

Thus, for many complicated reasons, cladistics is a difficult technique to apply well to the fossil record of primates. It is not a simple matter to decide which traits are relevant and to compile a list of them, nor is it patently obvious which resemblances are superficial and thus perhaps convergences and which indicate a common evolutionary pathway. As in most endeavors, intelligence and thought are required if things are to come out right.

Remember, too, that the list of monkey-ape differences that seems so simple—just tick off the traits and you know which sort of primate you have—is compiled from the traits of *modern* animals. The pull of the recent is at work here also. We have every reason to suppose that both monkeys and apes have evolved and changed during the millions of years since the last ancestor they shared in common. Therefore, although some of these traits arose at the time of the original divergence, some of them are of more recent origin and can't be expected in a monkey or an ape close to the time of their evolutionary split. Which traits are which? Only the fossil record can tell us.

To me *Proconsul* was clearly an ape, albeit a primitive one, or stem ape. Carol Ward's work on the locomotion of *Proconsul* proved that, 18 million years ago, apes had no tails and no ischial callosities or sitting pads. David Begun's work on the phalanges of *Proconsul* showed apelike adaptations for greater grasping ability in the thumb and big toe than are seen in monkeys. In my opinion, these features,

Hominoids

Aegyptopithecus *Victoriapithecus* Crown OWM ***Proconsul*** *Afropithecus* *Kenyapithecus* *Hylobates* *Sivapithecus* *Pongo* *Gorilla* *Pan* *Homo*

This cladogram shows the basal, or primitive, position of *Proconsul* when genera of various living and extinct hominoids are clustered according to their uniquely shared anatomical features. At the far left is *Aegyptopithecus*, an early stem anthropoid. The next lineage to diverge includes *Victoriapithecus* and other extinct and living Old World monkeys (Crown OWM). *Proconsul* and all fossile and living species to its right are hominoids. *Hylobates* is the living gibbon; *Pongo* is the living orangutan; *Gorilla* is the living gorilla; and *Pan* is the living chimpanzee. (© Alan Walker.)

combined with the dental and cranial characters, are enough to demarcate the origins of apes. For now, this seems to be the prevailing consensus among paleoanthropologists.

The alternative viewpoint, presented by Terry Harrison of New York University among others, is that none of the so-called Miocene apes are really apes at all. In his view, all of the Miocene apes I have discussed here should be considered primitive or stem catarrhines because of uncertainty in their relationships to modern apes. In other words, they are not *yet* apes; they belong to the more general group that later gave rise to Old World monkeys and apes. While I agree there is great uncertainty about the relationships among the

Miocene and modern hominoids, I think we have enough documented distinctions to consider that the large-bodied Miocene higher primates from Africa were apes.

The problem exemplified by *Simiolus* is that there is another group of fossil primates in the Miocene of Africa. These have often been called "small apes" because of their modest size relative to that of the great apes of today. *Simiolus* is among these "small apes," as are species such as *Dendropithecus, Kalepithecus, Limnopithecus,* and *Micropithecus,* all known from Kenyan fossils. Because they are small-bodied, and the lesser apes of today—the gibbons and siamangs—are also small-bodied, there has been a temptation to believe that some of these fossil species were the ancestors of the modern lesser apes, or hylobatids. But the only anatomical features shared by the living hylobatids and these "small apes" are primitive characters of the catarrhines as a whole; they can't be taken as evidence of a special, ancestor-descendant relationship. For now, these small fossil apes are perhaps best described as primitive or stem catarrhines. While Miocene hominoids such as *Proconsul* and *Afropithecus* were evolving specialized anatomy for living in the trees, the more primitive types of catarrhines persisted, sometimes at the very same sites.

Kenyapithecus is yet another Miocene ape that has contributed to the confusion, in part because of its complicated history of discovery and naming. The first species to be named, *Kenyapithecus wickeri,* was found at the 14-million-year-old site of Fort Ternan, in western Kenya, by Louis Leakey in 1962. In typical Leakey fashion, Louis was immediately certain that *Kenyapithecus* was both a direct human ancestor and a tool user, though the evidence for either interpretation was flimsy and has subsequently been discounted. In 1973, with Peter Andrews and Judy Harris (then Van Couvering), I led an excavation at Fort Ternan in hopes of finding more fossils of *Kenyapithecus wickeri* that would clarify its adaptations and relationships. We didn't find much of *Kenyapithecus,* though there were lots of fossil antelopes and giraffes and hippos and rodents and carnivores to keep us cheerful.

Fort Ternan does hold a special place in my memory, though. Sometimes when I give a talk about my fossil work, someone in the audience asks me what the best thing was that I ever discovered on a dig. I have to say it was something I found unexpectedly at Fort Ternan: the woman who is now my wife, Pat Shipman. She was then the graduate student of an old friend at another university, who had pressed me to allow her to come along on the dig. Reluctantly, I agreed, never anticipating the outcome. Digs can be remarkably romantic settings.

The *Kenyapithecus* story doesn't stop there. A similar animal called *Kenyapithecus africanus* was also found by the Japanese team at the Samburu Hills in Kenya and by an American team at Maboko Island in Lake Victoria, both of which preserve animals from about 15 million years ago. Controlled excavations on Maboko were started in 1987 by a husband-and-wife team, Brenda Benefit and Monte McCrossin, now at New Mexico State University. Their excavations have yielded many specimens of *K. africanus,* including a few limb bones. McCrossin's analysis of the postcrania from Maboko suggested that *Kenyapithecus* may have been semiterrestrial, but we need more and better specimens to verify this idea.

This was the state of affairs in July 1999, when the Japanese team led by Hidemi Ishida of Kyoto University announced that the fossils from Samburu should no longer be considered *Kenyapithecus africanus* but were a new hominoid species, *Nacholapithecus kerioi.* This reclassification was prompted by the discovery and analysis of a partial skeleton, which is unfortunately badly fragmented and somewhat distorted. Part of that skeleton of *Nacholapithecus* showed that the species had a long flexible spine and no tail, like *Proconsul nyanzae.* That was why they invited Carol Ward, Mark Teaford, and me to collaborate with them in reinvestigating the taillessness of *Proconsul nyanzae,* as I described earlier.

Only weeks later, before anyone had time to digest or evaluate the new information about *Nacholapithecus,* a team lead by Andrew Hill of Yale University and Steven Ward of the Northeast Ohio Universities College of Medicine announced another new genus and species of Miocene ape from Kenya: *Equatorius africanus.* This paper was based on a six-year study of an enigmatic partial skeleton of a hominoid that had been found by Boniface Kimeu, the son of my long-time collaborator Kamoya Kimeu, at Kipsaramon in the Rift Valley of Kenya.

Like the Samburu specimens that became *Nacholapithecus,* the Kipsaramon fossils had initially been taken to be *Kenyapithecus africanus.* While the new specimens from Kipsaramon closely resembled *K. africanus* from Maboko, the team decided that *K. africanus* plus the Kipsaramon specimens were different from *Kenyapithecus wickeri.* Remember that the name of a species or genus is attached to a particular specimen, known as the type or holotype; in this case, the type of both the genus *Kenyapithecus* and the species *K. wickeri* is a specimen from Fort Ternan. That means that if *K. africanus* from Maboko and Kipsaramon differ from the Fort Ternan type, the Maboko and Kipsaramon fossils have to be put into a new genus: *Equatorius.*

From the partial skeleton from Kipsaramon, *Equatorius* is estimated to be about the size of a modern-day baboon. It has thick-enameled teeth, a long flexible vertebral column, and limbs more adapted for ground dwelling and terrestrial locomotion than, for example, *Proconsul.*

It is a remarkable coincidence that two teams independently published their decisions to extract specimens from *Kenyapithecus africanus* and put them into new genera within a month of each other. Neither group knew the other was about to publish a similar argument. What this coincidence shows is how very poorly defined and known *K. africanus* was—and still is, if it exists at all.

Many scholars group *Nacholapithecus, Equatorius,* and *Afropithecus* into a subfamily, the Afropithecinae, because of the number of resemblances in their teeth and jaws. The curious thing is the differences among the limb bones of these genera. *Afropithecus* is interpreted as an animal that moved slowly through the trees on four feet; *Equatorius* is seen as a four-footed ground dweller; and *Nacholapithecus* was apparently clambering and arm-swinging through the trees. Can all of these interpretations be correct? Yes, and if they are, they would indicate a striking divergence of locomotor habits within three genera so closely related that they have been confused for one another.

Challenges to *Equatorius* came from two sources. One is Benefit and McCrossin, the researchers who work on the Maboko specimens. They agree that *Equatorius* and *Kenyapithecus* represent a single genus but they perceive a strong link between *Kenyapithecus africanus* and *K. wickeri* as well, which means that *Kenyapithecus* would be the appropriate genus for all three species. Another skeptic is David Begun, who believes that *Equatorius* is the same species as *Griphopithecus* from Germany and Turkey. If David is right, then by the rules of priority, *Equatorius* is invalid and the specimens from Kipsaramon ought to be called *Griphopithecus.*

What these confusions and renaming of specimens show is how hard it is to tell one of these Miocene hominoids from another. I know personally that a jaw with a few teeth is insufficient to distinguish *Afropithecus* from *Sivapithecus.* The renaming of *Nacholapithecus* and the debates over the proper relationships among *Kenyapithecus, Equatorius,* and *Griphopithecus* show that you need very good fossil specimens indeed to tell some of these Miocene hominoids apart from one another. The crucial differences may be in the limbs, hands, and feet, not in the more commonly preserved jaws and teeth.

One of the things we do know about Miocene apes, absolutely and for sure, is that there was an extraordinary abundance of them in Africa during the Miocene. We have no reason to believe that we have yet discovered all the Miocene apes that ever lived. What on earth was going on? Steve Ward, one of those who proposed the genus *Equatorius,* suffers from the same befuddlement at the excessive number of Miocene apes that I do. He is a congenitally irreverent expert in cranial anatomy as well as a keen observer of human foibles.

"What initially seemed to be a fairly nice, simple progression towards modern species got scuppered by the fossil record," he quips, and he is right.

When there were rather few Miocene apes, drawing hypothetical lines between the fossils and the living apes was easy. Some once thought little Miocene apes must be ancestors of hylobatids; the biggest ones must be the ancestors of gorillas; and everything in between probably led to chimpanzees. The trouble was those lines and the simple ideas behind them just weren't realistic. Even if you limit yourself to trying to think about only the African apes of the Miocene, the picture is one of an overabundance of species compared to any modern situation you can find. There are more Miocene hominoids still in Europe and Asia. The total number of distinctive species known today is twenty-something (the number can change any day with a new discovery), so sorting out relationships and adaptations becomes a truly formidable task.

While there was a stunning adaptive radiation of Miocene hominoids in Africa, there were amazingly few Miocene monkeys, and monkeys and apes rarely lived together. One of the most interesting exceptions is the monkey *Victoriapithecus macinnesi,* which apparently lived on ancient Maboko Island side by side with the primitive ape *Equatorius.*

Victoriapithecus is also known from a number of other Kenyan sites ranging in age from 19 million years to 12.5 million years, but its greatest abundance to date is at Maboko. At Maboko, *Victoriapithecus* is very common, making up about one in five of the thousands of fossil specimens recovered by Benefit and McCrossin during their excavations. Analysis of its postcranial bones shows that *Victoriapithecus* was a small monkey (6.6–11 pounds or 3–5 kilograms) with some terrestrial adaptations such as shortened fingers and toes and sitting pads, or ischial callosities, on its bottom. It is perhaps somewhat similar to the modern vervet monkey in this regard. In 1997, the Benefit-McCrossin team found a complete skull of *Victoriapithecus,* which qualifies as the earliest skull of an African

Miocene monkey, with an EQ about typical for a small-bodied monkey today.

This early monkey presents a primitive mix of traits from the two groups of monkeys that live today in the Old World, the beautiful leaf-eating colobines and the cercopithecines, such as vervets, baboons, and guenons. Benefit has argued effectively that the Victoriapithecidae was a taxonomic family that included the last common ancestor of all Old World monkeys. This means that *Victoriapithecus* and its close kin *Prohylobates* were stem Old World monkeys that lived before the colobines and cercopithecines evolved into separate groups.

Prohylobates is the only other monkey known from the African Miocene. It was originally thought to be an ape ancestral to the living gibbon, *Hylobates,* so it was given a generic name suggesting such a relationship. Though *Prohylobates* is now classified as a monkey, not an ape, and no one believes it is ancestral to *Hylobates,* the name is technically valid and thus persists. A few specimens of *Prohylobates* come from Buluk along with the hominoid *Afropithecus;* other researchers have found *Prohylobates* at the north African sites of Wadi Moghara, Egypt, and Gebel Zelten in Libya. In the parts we know so far—broken jaws and teeth—*Prohylobates* generally resembles *Victoriapithecus,* but we have much yet to learn. Once we have more body parts, the story may be rather different.

Monkeys were genuinely rare in Africa during the Miocene, which suggests that they may have specialized in a habitat that was uncommon. Many years ago, John Napier noticed that monkeys as a group seem to show a great many adaptations for terrestriality, even though many monkeys are now arboreal. John may have hit upon the evolutionary solution to the disjunct distribution of many Miocene apes and Miocene monkeys: perhaps monkeys were specialists in savannah or open woodland.

How does this hypothesis fare when tested by what we know of the sites where Miocene hominoids were living? Some of the best-known sites are those where *Proconsul* has been found and where monkeys have not: Rusinga and Mfangano, which have yielded huge fossil collections. Tens of thousands of fossil species from insects to lizards and mammals show adaptations to ancient rain forests. We even have the preserved nuts, fruits, and tree trunks of those forests on Rusinga and Mfangano, such as the tree in which the *Proconsul* skeleton was found. The leaf beds on Rusinga prove that many of the leaves have drip-tips, the elongated points that help water run off of leaves in tropical rain forests. Other Miocene ape sites, though less well known, were also probably forested, suggesting that many of the ecological niches in forests that are now filled by monkeys were exploited by the Miocene apes.

From their work on Maboko, Benefit and McCrossin hypothesize that the earliest monkeys specialized in habitats where the forest, open woodland, and savannah met in a complex mosaic; they also hypothesize that the early monkeys were eating hard fruits and nuts that grew close to or fell to the ground. This idea can be tested when more information is available on the paleoecology of sites where fossils of both monkeys and apes are present.

We know that the environment has changed dramatically in Africa between the Miocene and modern times. In the early Miocene, African forests formed a broad, continuous belt across the continent. As volcanoes rose and erupted, especially along the East African rift valley, lava flows resculpted the landscape and changed the rainfall and drainage patterns, and ash falls altered the chemistry of the soils. Of course, these changes did not happen everywhere in Africa, but the fragmentation and retreat of the forest in East Africa probably contributed to the extinction of many of the Miocene ape species we now know about. When much of Africa was forested, apes were common and monkeys were rare. Now the forest has shrunk and apes are rarer, monkeys more common. As acid rain, global warming, agriculture, and the insatiable human need for fuel interact, the forest continues to diminish on a daily basis. Apes are becoming steadily rarer while monkeys still thrive, but for how much longer?

When we look at a graph of the fossil record of higher primates

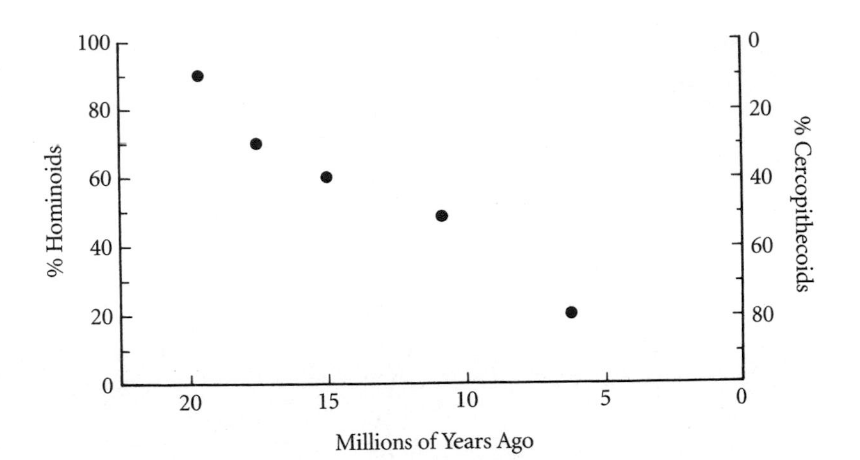

Ape species (hominoids) were numerous in the Miocene and declined through time, while Old World monkeys (cercopithecoids) show the opposite trend. Various species of *Proconsul* were part of the Miocene ape radiation. (Adapted with permission from J. Fleagle, *Primate Adaptation and Evolution* [New York: Academic Press, 1999], 506.)

in Africa from the Miocene to the present day, we can see just what has happened. In the earliest period, there were nine species of hominoid, six stem catarrhines, or "small apes," and four monkey species. The proportion of hominoids drops through the next four periods, from nine hominoids to seven to one, while the number of monkeys stays steady at four species. Then, in the period from 5 million years ago until 10,000 years ago, there are no fossil hominoids in Africa (though the ancestors of chimpanzees and gorillas were almost certainly present), no stem catarrhines, and thirty species of monkey. If we add in the living primates of Africa, then the total rises to two genera and four species of apes (mountain and lowland gorillas, common chimpanzees, and bonobos) and sixty-one monkey species in eleven genera. From about 23.5 million years ago until 12 million years ago, apes flourished and diversified in Africa; then their numbers declined dramatically.

In Europe (including the Middle East) and Asia, the abundance of primitive apes came slightly later, from about 16 to about 8 million years ago, after which time those apes too began to disappear.

Why? How?

One of the people with an interesting perspective on this evolutionary muddle is Mike Rose of the New Jersey School of Medicine. Mike and I have a long history together. We first met while we were doing national service in the Royal Air Force. We were probably the only two men on the Hereford (England) Royal Air Force base who brought scientific books along as spare-time reading. After serving our tours of duty, we both went to Cambridge University, Mike to study botany while I studied geology and paleontology. I then went on to the Royal Free Hospital in London to study anatomy with John Napier, while Mike went to the Middlesex Hospital for medical training. After qualifying as a physician, Mike decided that anatomy, not doctoring, was where his greatest strength lay, so he followed me in joining John's group. In 1965, I left London to take a job teaching anatomy in Uganda while I finished up the study of lemur limb bones and locomotion for my Ph.D. thesis. Soon Mike came out to Uganda to do the research on monkey locomotion that formed the basis of his Ph.D. thesis. Together, we excavated a few fossil sites in Uganda and described the Moroto vertebrae. In 1969, I moved to Kenya to teach anatomy at the newly formed University of Nairobi Medical School in Kenya; Mike soon followed. In 1973, I left Kenya to take up a job in the United States at Harvard University; a few years later, Mike left Kenya to take up a job at Yale University. It became a standing joke between us that, wherever I went, Mike was soon to follow. Mike's wife, Cordelia, is an exceptional and delightful woman who once asked me if she could have approval over my job offers, so that she could have some say in where they were going to live next.

Mike is a quietly brilliant analyst who has studied the limb bones of nearly every Miocene ape in the world, as well as those of many monkeys. He is also an aficionado of science fiction and African music. Once, during a conference on Miocene primates at the American Museum of Natural History, he turned to my wife, Pat, and said,

"When I look at the postcranial bones from the Miocene apes, I get a fairly consistent pattern from many species, but it is nothing like what we see in modern apes. Maybe we should consider the ones that survived as the bizarre ones."

Mike's insight is worth considering. What if we invert our prejudices and take the Miocene as "normal"? Then, we would clearly see the modern numbers and distributions of apes as the depauperate remnants of a former abundance. The apes we know today, from this perspective, are the exceptions, the strange survivors who have not—yet—been forced into extinction by the changed world in which they live.

10

Something to Chew On

In the 1980s, I learned that scholars from various disciplines were developing new ideas about the life-history strategies of different species and new ways to detect them. Here was some information that was truly biological and exciting.

By "life-history strategies," I mean the timing and duration of the developmental and reproductive events in a species—how an animal grows and matures and how it reproduces—over the course of its lifespan. If I could find clear evidence of the life-history strategy used by *Proconsul,* and direct evidence of only one of these events, then I could deduce a great deal more. The major events in a species' life history are often tightly linked to one another.

What kinds of events was I hoping to learn about? Perhaps I would be able to learn how long gestation lasted in *Proconsul* or what the usual spacing between births to a single mother was. If I could find out when a *Proconsul* was typically weaned, that would in turn imply something about how fast the members of that species matured. If I knew when physiological infancy ended for *Proconsul* in terms of the number of days or weeks since birth, then I could make a good guess about when sexual maturity or adulthood arrived in this extinct species or how long individuals lived, on average.

Underlying the concept of life-history strategy are some very basic biological principles. If you review information about a very wide range of mammalian species from mouse to elephant, the timing of these key life events correlates closely with both body size and brain size. In other words, body size or brain size can be used to predict the timing and length of gestation or age at sexual maturity in an extinct animal. Of course, there are exceptions to any general biological rule like this one—and sometimes the exceptions are the most interesting.

Paleontologists can only very rarely observe events such as birth—for example, there is a fossilized ichthyosaur that died giving birth—or sexual maturity in extinct species. We have to rely upon the relationship between these life events and the individual's chronological age. Chronological age can't be *directly* observed on fossils either, although some of the *effects* of age can be observed on the bones. What can be observed is which teeth have erupted through the gums and become worn through use. Fortunately, teeth in any species erupt in a fairly consistent and predictable sequence. Most parents know that human babies almost always erupt their deciduous, or milk, incisors first, for example, and only much later (at age five or six) erupt their permanent molars. Eventually all milk teeth fall out and are replaced by permanent, or adult, teeth. Although the sequence of dental eruption is not identical in all mammals or even in all primates, if you know which teeth have erupted and come into wear in a fossil specimen, you can figure out how old that individual might have been when it died. This statistic is referred to as *age at death* to distinguish it from geological age in years before the present, or *antiquity.*

In mammals, the time when the first permanent molar comes into wear is a key event marking the end of a physiological stage known as infancy. Understand, though, that infancy as defined in this sense is not necessarily the same thing as infancy in common parlance. In nonhuman primates, infancy ends and the subsequent juvenile phase begins at the time of weaning, which is also close to the time when wear begins on the first permanent molars (M1s). This is intuitively obvious because, once weaning is completed, the young primate must start eating adult food or starve. The molar teeth are usually the ones that break the food up into small chunks or a mush that can be swallowed. For example, in chimpanzees, M1 eruption and the beginning of the juvenile period occurs at 39 or 40 months. Measuring age as the number of months since birth—rather than saying "3 years and 3 or 4 months"—makes it easier to compare animals of different sizes and rates of development. In contrast, macaques are weaned and their M1s erupt at about 18 months after birth.

In humans, developmental or physiological infancy ends between 64 and 70 months in various populations. This, however, is one of those fascinating deviations from the prediction, because in humans, weaning usually occurs well before the end of developmental infancy. Very few human mothers nurse their children until the age of 64 months, whether they live in industrialized nations or in the developing world. Weaning has occurred—and socially defined infancy has ended—long before 64 months.

So how and why does the human species wean infants unusually early in their physiological development compared to other mammals? One reason is that humans *can* and other mammals *can't*. Humans have weaning foods, such as the milk of other species or soft and mushy cereals, which can provide the youngster with adequate nourishment without exhausting the mother's ability to produce breast milk. The invention of weaning foods must have made a remarkable difference in our evolutionary success. Weaning foods freed human mothers from the nutritional tyranny of breastfeeding their young until the end of physiological infancy. Mothers who have access to weaning foods can resume their normal activities—such as gathering or growing food or hunting—or get pregnant again much sooner after giving birth than females in other species that have similarly helpless and dependent infants. What this means is that human mothers are able to make more babies during their reproductive lifetime and have more of them survive to reproductive age, simply because of the invention of weaning foods. Having more babies that survive better is a fundamental aspect of evolutionary success. Nonhuman primates don't have that option, so mothers must nurse their offspring until they have enough teeth and jaws powerful enough to tackle adult food. As a result, humans can and do reproduce faster than would be expected for another primate of the same body size.

Weaning infants early in their developmental history is not the only peculiarity of the human life-history strategy. Compared to monkeys, both apes and humans deviate oddly in their life-history strategies. Apes have slowed down their developmental maturation relative to that of monkeys, stretching all of the life segments out like an elongated rubber band.

Humans have done this and more, for all phases of the human life history are wildly out of kilter with those of other primates. Weaning may correspond roughly with the end of what we call infancy in humans, but newly weaned children are still in a protected and highly dependent phase of life that is effectively a type of infancy. Humans also stretch the juvenile period out to include adolescence, which is a peculiarly human invention, as Barry Bogin has pointed out. Juveniles—preadolescent humans—are unusually small and appear immature compared to apes at the same physiological stage of development. In nonhuman primates, sexual maturity comes at the end of the juvenile period, a point that is marked by the emergence of the second molar.

A human also ends his or her juvenile period with the emergence of the second molar (M2), which is about the time when the sex hormones kick in, between 11 and 13 years in most populations. Unlike apes at sexual maturity, these young humans are neither quite adults nor still juveniles. They are or are about to be reproductively mature, but their relatively small size and apparent youth conceals their sexual maturity. Adolescent humans often continue to have the social status of juveniles and live with their natal families. Bogin hypothesizes that human adolescence is a special phase, an evolutionary mechanism designed to mask sexual maturity and prolong the period before the first reproduction. Why would any animal want to do that? Bogin's answer is that adolescence serves to extend childhood in order to prolong the period of learning within the family. Humans have a lot to learn about all sorts of things before they are ready to function as full adults. And, unlike other primates, humans often undergo a distinct growth spurt toward the end of adolescence. This period of unusually rapid growth—when kids seem to outgrow the clothes that were too long for them only last week—allows them to catch up to adult size. Often, humans assume a sexually mature role soon after they begin to look like adults. Even adulthood is carried to extremes in humans, who routinely live well past the end of their reproductive years, unlike most other mammals.

One of the pioneers in the study of life-history strategies in primates is Holly Smith, a friend of mine at the University of Michigan. During her career, Holly has examined literally thousands of dental radiographs of immature monkeys, apes, and humans to learn how to estimate the age at death of any particular skeleton. She has refined dental aging techniques considerably. For example, a child with its first permanent molar in wear and a second molar that is developing but not yet in wear is older than 64–70 months (the average age for eruption of M1) but younger than 132–144 months (the average age for eruption of M2). By assessing how complete the root or crown of each tooth in the skull or jaw is, Holly may be able to arrive at a more precise estimated age of perhaps 98 months.

The trick is to transform a developmental age, based on precise observations, into months since birth. Holly has compiled a large database on the dental development of the teeth of primates of known chronological, or calendrical, age. From this she has created models of how dental age correlates with calendrical age in monkeys, apes, and humans. When confronted with specimens of an extinct species of higher primate, Holly must first decide whether it grew like a monkey, like an ape, or like a human and then, using her models, she can estimate the fossil's age at death.

In broad terms, there are two divergent life-history strategies that form the ends of a continuous spectrum. There are animals that "live fast and die young," to borrow a phrase that Holly uses often. Typically, these animals mature quickly and have lots of babies that grow up rapidly with little parental care. The emphasis is on quantity over quality. These are the mice on the proverbial mouse to elephant curve. At the opposite end of the spectrum are animals who "live slow and die old," like humans—or elephants. It is if the tune to which humans dance is set at a slow tempo. Such species have few babies but invest in each one heavily in terms of parental care and teaching. This is the quality over quantity approach. All primates fall towards the slow, or elephant, end of the spectrum, though humans are extreme, so you might think that there was nothing more to know. But you'd be wrong. There is a lot of variation within higher primates in terms of the tempo of their development.

In the course of her work, Holly has emphasized that being able to predict the timing of life-history events *on average* tells you very little about any particular individual animal. This is a type of error similar to using the average height of a modern human to predict the height of any particular individual; you are more likely to be wrong than right. And, of course, the really interesting species are the ones that don't do what you'd expect for a creature of a particular body size. Then you have to puzzle over why this one species has deviated from the norm and what evolutionary benefit might result from that deviation.

Jay Kelley pointed out that one of the distinguishing characteristics between monkeys and apes is a shift in life-history strategies. This sort of change is a called a grade shift because it occurs not between two individual species, but between two groups of species (in this case, monkeys on the one hand and apes on the other) that differ in their grade of evolutionary development. As a general rule, all great apes grow more slowly than all monkeys, and therefore apes spread out all the developmental and reproductive events in their life histories over a longer period of time. This difference means that the debate over whether the fossil Miocene species were already apes or were stem catarrhines (monkeys) could potentially be resolved by life-history data. If the Miocene species had slow, extended life histories, then they were already in a fundamental sense apes.

In same paper, Jay also estimated the time of eruption of the first molar in *Proconsul heseloni* from its estimated brain size: about 19–20 months. When he compared the fossil ape to monkeys of similar body size (the rhesus monkey, the pig-tailed macaque, and the crab-eating macaque), *P. heseloni* took about 23 percent longer to erupt its M1s (according to his estimate) than these monkeys. Thus his estimate indicates a slower maturational rate in *Proconsul* than that observed in Old World monkeys. However, compared to a modern ape—a chimpanzee—*P. heseloni* reached the same developmental marker in half the number of months, or twice as fast. So was *P. heseloni* living like a monkey? No, but it wasn't yet living like a modern ape either. Yet Jay's work suggested that *P. heseloni* had already begun to deviate from the monkey standard in the direction of the apes.

Though Jay's work was a reasonable first approximation, and he was cautious in interpreting the significance of his results, I was uneasy about the possible errors involved in this estimate-based-on-an-estimate. What I wanted to know was *exactly* how long *Proconsul* took to grow up, not how long *Proconsul* took to grow up if our estimate of brain size was correct and if *Proconsul* did not deviate widely from the general rule of the relationship between brain size and developmental tempo. I needed to find a way to determine the age of an individual specimen with precision.

Fortunately, I soon became aware of some techniques being developed in the 1980s that might be able to tell me how *Proconsul* grew and how rapidly it grew. This elegant new approach relied upon information about how teeth grow at the cellular level. One of the leaders in this field, Christopher Dean of University College, London, sums up the basic idea simply: "Teeth carry their past in their internal architecture." Not many people realize that the precise details of the formation of every single tooth are preserved forever in its microscopic structure. It is an observation of stunning simplicity and wonderful importance.

Teeth are made up of two basic substances. Enamel is the hard, white tissue on the outside of teeth. Enamel is comprised of a series of calcified prisms, which are laid down in a complex pattern, and the outer surface of the tooth is covered by an amorphous layer of enamel known as the cuticle. Dentine is the underlying, slightly yellowish core of each crown, which is usually visible only in the tooth root or in broken or heavily worn tooth crowns.

These substances—enamel and dentine—begin forming in the embryo from two layers of cells known as ameloblasts (enamel-forming cells) and odontoblasts (dentine-forming cells). Each of these tissue-forming cells secretes its own organic matrix, which serves as the framework upon which the adult tissue is formed. After the cells have laid down the matrix, crystals of hydroxyapatite are deposited within the matrix in a process known as mineralization. The mineralization continues, hour by hour, day by day, and month by month until the tooth is completely formed. With the exception of minor repairs of cracks in the pulp cavity through the growth of secondary dentine, a primate tooth crown can grow no larger once it has erupted through the gum because the ameloblasts on the outside of the tooth are shed and die upon eruption.

A third hard substance, cementum, and dental ligaments anchor the tooth root into the bony socket. Cementum grows much more slowly than either enamel or dentine, but often you can see seasonal bands in cementum as the tooth erupts that correspond to increases or decreases of deposition like those in tree rings.

What is astonishing is that, by sectioning a tooth very thinly, grinding it even thinner, and inspecting the slice under a microscope, you can actually see the fine increments of mineral that were deposited in a single day. In sectioned enamel, these increments are known as cross-striations; they look like fine lines running perpendicular to the main axis of the enamel prisms. Cross-striations known as von Ebner's lines are also visible in histological sections of dentine. Like many biological processes, the mineralization of enamel and dentine runs on a biological clock, a 24-hour cycle known as the circadian rhythm. Proving that these cross-striations represented a day's interval of mineralization took a great deal of hard and painstaking work by a number of dedicated researchers. Once that was completed, a new technique for investigating dental formation was in hand. If you have patience, formidable technical skills, and luck, you can make a section of a tooth and count the exact number of days over which that particular tooth crown or tooth root formed.

There are higher-order divisions within these hard tissues too, which reflect a longer cycle that occurs in increments of several days, known as long-period striae, or striae of Retzius. No one knows why these long-period striae form or what physiological process is thus recorded. No one knows what is signified, biologically, by having a longer or shorter interval between long-period striae. But these long-period striae are readily observable features that are very useful in calculating chronological age. Where they encounter the tooth's surface, they can be seen as a series of regular ridges and troughs that encircle the tooth's circumference. These surface features are called perikymata. On an unworn and well-preserved tooth, you can count perikymata to calculate the time the formation of that tooth took.

Fortunately, teeth are hard tissues that not only record their own process of growth and development but that also often survive the natural destructive processes that occur after death and thus become fossils more often than other body parts. Unfortunately, counting the perikymata visible on the surface of a fossilized tooth does not immediately give you the number of days, weeks, or months during which that tooth was formed. To find out how many days one long-period stria, or perikymatum, represents, you need to establish the periodicity of these long-term markers in that particular species. The question is: How many days does it take to form one stria in this species? To find this out, you have to section several teeth and count the daily cross-striae on several individuals to determine the average number of days between one long-period line and the next. From

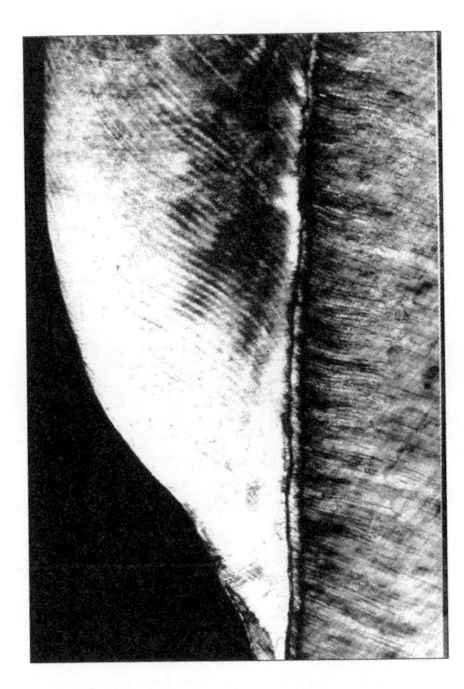

This micrograph shows a histological section through a *Proconsul* tooth. The junction between the inner dentine and outer enamel of the tooth appears to be a clear, nearly vertical line on this image. At the upper left, the intersections of the enamel prisms (running horizontally) and the long-period striations (running diagonally lower right to upper left) are clear. (© M. C. Dean; reprinted with permission from Elsevier from A. D. Beynon, M. C. Dean, M. G. Leakey, D. J. Reid, and A. Walker, "Comparative Dental Development and Microstructure of *Proconsul* Teeth from Rusinga Island, Kenya," *Journal of Human Evolution* 35 [1998]: 196.)

studies of living monkeys, apes, and humans, we know that long-period striae may represent intervals anywhere from 4–5 days up to 11–12 days. The same studies showed that a species has a consistent, average periodicity to its long-period striae, though there is some variation between individuals. Nevertheless, once you have taken sections from several individuals and determined the average periodicity of perikymata for that species, you can use that number over and over again in calculations for different teeth.

There are also singular, traumatic events or periods of time that leave dental traces. The most common and most important is a neonatal line, formed by the trauma of birth. Another line may indicate the onset of weaning and yet others may reflect serious illnesses or physiological stresses such as periods of starvation or malnutrition. Identifying the neonatal line is especially important if, as was the case with the *Proconsul* study, you hope to determine the number of months that have passed *since birth* when a particular tooth's crown or root was completed.

Why these minutiae are interesting is because of the differences between individuals, sexes, and species. A question: As the apes evolved a slower life-history strategy, did they continue to form their teeth at the same incremental rate as the "faster" monkeys, with pauses between teeth, or did they slow down the entire process of secretion and mineralization? This was one of the questions that Chris Dean, working with Bernard Wood, now of George Washington University, was able to answer.

Chris's trajectory as a scientist shows how an agile mind can spring from one field to another, even though an outsider might view them as totally disparate. Chris trained as a dentist, enjoying the combination of technical and artistic skills that dentistry requires. He has a keenly inquiring mind—despite the fact he never felt he was an outstanding student—and an ability to burrow into tiny details and follow them through. He likes to make things and do things with his hands, such as growing vegetables on his allotment, restoring grand pianos, or remodeling the row house in London he lived in for years.

In the early part of his career, Chris treated "schoolsfull of children" before he realized that clinical dentistry would not hold his interest forever. Seeking a greater intellectual challenge, he took a master's degree in human biology with the anthropologist Geoffrey Harrison at Oxford, where he was exposed to the mysteries and big questions of human evolution that he found so lacking in the daily practice of dentistry. Before long, Chris had enrolled in a Ph.D. program with Bernard Wood, then at the Middlesex Hospital in London, where he began to research how the base of the skull grows and changes through life. To support his studies, he continued his clinical practice part-time. Chris's research project foundered temporarily when he found out how inadequate the available techniques for determining age were. He couldn't study growth in the skull without knowing the age at death of each of the skulls he was examining. Given his intensive background in dental anatomy, perhaps the most obvious solution to his frustration was to construct his own aging scale by documenting the fine details of dental development in great apes and humans. In this aspect, his work closely paralleled some of what Holly Smith did. The difference was that while Holly used radiographs to see how complete the major structures (crown or cusp and root) of each tooth were, Chris looked in close detail at the individual growth patterns recorded within the histological structure of each tooth.

One of the most surprising findings of his work was that apes and humans spent the same number of days growing each individual tooth. If that is true, Chris wondered, why does a great ape complete all its dental development by 144 months (12 years) when it takes a human 216 months (18 years) to do the same thing? Apes and humans have the same number and kinds of teeth: in each quadrant of the mouth, both have a central and a lateral incisor, a canine, two premolars, and three molars. Thus he knew that humans didn't take longer to complete their dental development because they had extra teeth.

What other explanation was there for the elongated human development? Chris had noticed that, in both humans and great apes, each molar in the mouth begins to mineralize before the previous one has completed its mineralization. Thus the second molar (M2) begins to mineralize before the first (M1) is completed and the third (M3) begins to mineralize before the second (M2) is completed. When he documented this phenomenon in detail, Chris found that there is less chronological overlap between the mineralization of adjacent teeth in humans, so the whole sequence takes longer. If a genetic mutation could occur once to adjust the extent of overlap between adjacent teeth, then it could have occurred twice. The same mechanism might have caused the difference in developmental tempo between monkeys and apes, with monkeys developing faster because they had more overlap in the mineralization of adjacent teeth than apes do. It made beautiful sense.

After completing his Ph.D., Chris did another stint of clinical practice before he got reinvolved in the world of research through meeting Tim Bromage, then a graduate student at the University of Toronto. Tim was investigating the periodicity of perikymata in early human ancestors and soon he and Chris were collaborating on the project. They were able to document the typical time intervals between perikymata on a number of early species in the human family. At about the same time, Chris also met the dental histologist David Beynon and his technician, Donald Reid, from the Dental School of the University of Newcastle-upon-Tyne. That trio embarked on a longer collaboration to try to find still better ways to determine the age at death of individual fossils from the microscopic anatomy of teeth. As the collaboration developed, their personalities jelled in that special way that happens sometimes when you get very lucky. Chris was the font of ideas, David was the expert histologist, and Don was the brilliant man at the microscope. Don is one of those people with an extraordinary gift for microscopy and he loves it.

"Once you sit down at the microscope," Don once told me, "and you've got a wonderful ground section—it doesn't matter whether it is a modern human tooth or a fossil one. You can see things in it, you can see patterns, you can see life in them. It's the excitement of trying to glean out what's going on that makes it all worthwhile."

When I asked Chris if he, David, and Don would be interested in looking at the histology and growth of *Proconsul* teeth, I got a "yes, please!" right away. I wanted to know how many days old a *Proconsul* was when its first and second molars came into wear, and when *Proconsul* reached physiological adulthood compared to apes and monkeys. I knew they could tell me if all *Proconsul* species grew at the same rate and whether they had monkeylike, apelike, or entirely novel life-history strategies. Along the way, this triumvirate discovered some surprises.

Fortunately, we had lots of teeth of *Proconsul*, many of which were not in place in jaws so no bone would be destroyed in making sections. (Museum curators tend to be very reluctant to let specimens in their care be subjected to destructive techniques, and rightly so.)

Moreover, Chris and his collaborators had perfected a marvelous technique for taking very thin slices of fossil teeth, using an annular diamond saw to remove a 100-micron-thick slice of tooth. One hundred microns is about the thickness of a single human hair. As a precaution, in case the specimen was irretrievably damaged during sectioning, Chris would first make a negative mold of the tooth, using an impression material that recorded even microscopic features accurately. Then from the mold he made a positive cast that could be left at the museum that held the specimens to record the details of the tooth before the sectioning occurred. After removing the slice, Chris would glue the two halves of the original specimen back together. Only researchers with the keenest eyesight could tell that a section had been removed.

The study involved taking sections from eleven teeth of *Proconsul heseloni*—most of them from one of the juvenile individuals and a few from an adult individual from the Kaswanga Primate Site—and two of *P. nyanzae*. The team combined the information on growth and life history from the sections with data on the number of perikymata visible on the surfaces of forty-four *Proconsul* teeth (including *P. heseloni, P. nyanzae,* and *P. major*) that were stored in the National Museums of Kenya in Nairobi. Back home in England, Don ground the sections to exquisite thinness, so light would shine through them and reveal the intricate histology. Cross-striations and long-period striae were painstakingly counted, recounted, and measured, with the aid of both the microscope and photomontages of single teeth.

Even though there is a peculiar beauty to the histological structure of teeth—one that Chris and Don both treasure—this sort of work is incredibly time-consuming and requires the utmost concentration. The whole task is greatly complicated by the fact that the enamel prisms do not run straight but interweave in a complicated pattern, so following a single enamel prism throughout its length is impossible. In addition, taking sections in comparable planes in different teeth is tricky. Finally, if a section is broken or damaged during preparation, there is no possibility of taking another. It is difficult enough to persuade a curator to let you take a single, very, very thin section, without coming back to make more because you accidentally dropped your specimen or ground it so thin that it shattered.

From sections of the fossils, Chris, David, and Don were able to show that the crown of the first molar in *Proconsul heseloni* was completed after 15 months of growth, when the individual was about 14 months old. In other words, the crown of the M1 began forming about a month before birth, as judged from the placement of the neonatal line, and continued for 14 months after birth. How much later did the M1 erupt into wear? The team observed that their results on crown completion of M1 are "not incompatible with" Jay's estimate of 19–20 months for eruption. They knew from their sections that root formation occurred at a rate of 6.4 microns per day in *P. heseloni.* If M1 erupted at about 19–20 months, then the first molar of *Proconsul* had only about 1.25 to 1.5 millimeters of root when it started chewing food. This seems to me like too little root to make the tooth functional, and Jay and Chris agree, although we are all using our intuition rather than any hard data. It seems more likely that M1 stayed in the gum until 2–4 millimeters of root were present, yielding an eruption time of 24–26 months, which would coincide with the onset of weaning. Therefore, we can say with certainty that the youngest individual from the Kaswanga Primate Site died before weaning, because it was only about 1 year old.

The crown of M2 of *P. heseloni* was completed at about 27 months and the crown of M3 was completed at about 43.5 months of age. Adulthood, as indicated by the completion of both the crown and root of the third molar in *P. heseloni,* occurred at about 78 months.

Because a neonatal line and another stress line, perhaps caused by an illness, are present on each of the isolated juvenile teeth of *P. heseloni,* Chris and his colleagues were able to confirm that all of them came from a single individual except the M2, which lacked the second line. Thus they could make a chart of dental development in

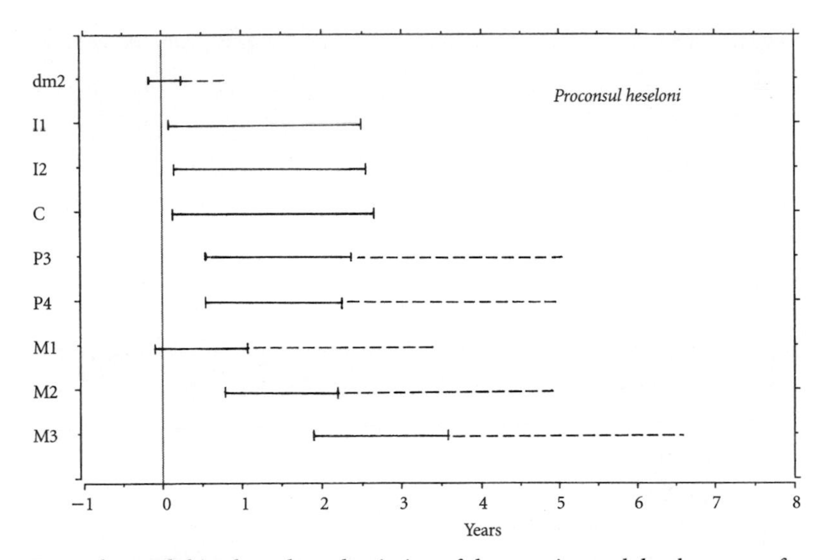

From the tooth histology data, the timing of the eruption and development of *Proconsul heseloni* teeth was established. Abbreviations: dm2 = deciduous second molar; I1= first incisor; I2 = second incisor; C = canine; P3 = first premolar; P4 = second premolar; M1, M2, M3 = first, second, and third molars. The vertical line represents the time of birth where it is known. Solid lines represent crown formation; dashed lines represent root formation. (Reprinted with permission from Elsevier from A. D. Beynon, M. C. Dean, M. G. Leakey, D. J. Reid, and A. Walker, "Comparative Dental Development and Microstructure of *Proconsul* Teeth from Rusinga Island, Kenya," *Journal of Human Evolution* 35 [1998]: 200.)

P. heseloni incorporating both calendrical and developmental age.

Unfortunately, the same cannot be done for *Proconsul nyanzae* because the British team hasn't yet sectioned enough of its teeth. They can show that *P. nyanzae* grew more slowly in absolute terms than *P. heseloni,* because M1 and M2 took more months to form in the larger species than in the smaller.

The same techniques have been picked up and used by Wendy Dirks of Emory University to enlarge the number of species for which such information is available. She chose to examine the teeth

The age (in months) of molar crown completion in various primates. The values in the table are average ages based on upper and lower molars.

Species	M1	M2	M3
Proconsul heseloni	14	27	43.5
Hylobates lar (gibbon)	13	28.6	52♀–60♂
Symphalangus syndactylus (siamang)	13.7	28.8	42
Papio hamadryas (hamadryas baboon)	16.4♀	36♀	68.5♀
Semnopithecus entellus (langur)	10.7	29.4	44.8
Pan troglodytes (chimpanzee)	30	60	90

Sources: Data on *P. heseloni* are from A. D. Beynon, M. C. Dean, M. G. Leakey, D. J. Reid, and A. Walker, "Comparative Dental Development and Microstructure of *Proconsul* Teeth from Rusinga Island, Kenya," *Journal of Human Evolution* 35 (1998): 163–209; data on living species are from W. Dirks, D. J. Reid, C. J. Jolly, J. E. Phillips-Conroy, and F. E. Brett, "Out of the Mouths of Baboons: Stress, Life History, and Dental Development in the Awash National Park Hybrid Zone, *American Journal of Physical Anthropology* 118 (2002): 239–252, and W. Dirks, "The Effect of Diet on Dental Development in Four Species of Catarrhine Primates," *American Journal of Primatology* 61 (2003): 29–40; data on *Pan* are from D. J. Reid, G. Schwartz, M. C. Dean, and M. Chandraseka, "A Histological Reconstruction of Dental Development in the Common Chimpanzee, *Pan troglodytes*," *Journal of Human Evolution* 35 (1998): 427–448.

of lesser apes (gibbons and siamangs) and two monkey species (hamadryas baboons and langur monkeys) that are roughly similar in size to *P. heseloni*. Siamangs and langurs are close in size to *Proconsul heseloni*—about 20–24 pounds, or 9–10 kilograms—while the gibbon has a body mass of only about 12 pounds (5.34 kilograms). Females of hamadryas baboons are also about 24 pounds (11 kilograms) but males are twice as large, so Wendy used only females in her study. The general similarity of body size makes it reasonable to compare the age at crown completion among these five species.

At the bottom of the table on page 216 displaying Wendy's data, I have added information on chimpanzee development based on British work, but it is important to remember that chimpanzees weigh about three times as much as *P. heseloni* and the other species included in this table.

Of the animals that weigh 20–24 pounds (9–11 kilograms), the siamang and *P. heseloni* are remarkably similar in terms of the age at which they complete their molar crowns, regardless of which molar you consider. Let's take those as representing a normal pattern for dental development in a small-bodied higher primate. The effect of larger body size on slowing down dental development can be seen by comparing this small-bodied ape pattern to the chimpanzee pattern at the bottom of the table. The chimpanzee is much larger and, as expected, dental development takes much longer.

What about the other species in Wendy's study? The gibbon shows the same small-bodied pattern as the siamang and *Proconsul* in terms of the completion of M1 and M2, but the completion of M3 seems to be delayed. The langur follows the normal pattern in terms of completion of M2 and M3 but M1 is completed very early. The real outlier in this group is the hamadryas baboon, which completes all of its molars later than the other small-bodied species in this table. Slowed development is supposed to be an ape pattern, not a monkey pattern, or perhaps a large-bodied pattern and not a small-bodied pattern. However, a baboon is indubitably a monkey and female hamadryas baboons are not appreciably larger in body size than langurs, gibbons, and siamangs. What is going on here?

Possibly the slowed development of the baboon teeth is due to the fact that males of the same species are much larger-bodied, even though Wendy sectioned only female teeth. Wendy suggests that there is yet another potent influence on dental developmental in addition to body size: diet. A number of different studies have found that animals which consume a higher proportion of leaves (folivores) tend to erupt their molar teeth earlier and develop them more rapidly than animals which eat more fruit (frugivores). Leaves tend to require a lot of chewing and digestion time to yield adequate nutrients and young, tender leaves tend to emerge all at once in seasonal environments. Fruit is softer and more readily digestible and trees produce fruit over a longer period. These differences in the availability and mechanical properties of these two important dietary elements has led researchers to hypothesize that there is an evolutionary pressure for young folivores to get their teeth working faster—a pressure not felt so keenly by young frugivores. And even though these are broad generalizations, since primates are rarely exclusive folivores or exclusive frugivores, the underlying distinctions are real.

To explore the hypothesis that diet affects dental development further, Wendy collected the data on two pairs of closely related species presented in the table above. She first chose a pair of monkeys with different diets: the langur is more folivorous and the hamadryas baboon is less folivorous, eating more fruit. She also selected a pair of lesser apes, in which the siamang is more folivorous and the gibbon is more frugivorous. As the hypothesis predicted, she found that the more folivorous member of each pair had faster dental development. What is especially interesting is that the dietary effect is strong enough to erase some of the difference between the so-called rapid dental development of monkeys and the so-called slow dental development of apes.

Do the resemblances between the dental development of the more folivorous siamang and langur and the development of *Proconsul heseloni* mean that the fossil primate ate leaves? Probably not. The teeth of *P. heseloni* are shaped like those of a classic frugivore among primates, with broad central incisors and molars with low, rounded cusps. In contrast, primate folivores tend to have narrower incisors and sharp-crested, high-cusped molars that are good at chopping up leaves. Further, different dietary substances produce different wear patterns on dental enamel and *Proconsul* wore its teeth like a frugivore and not like a folivore. Finally, my personal conviction that *Proconsul* was a frugivore is bolstered by another, remarkable piece of evidence. One of the special sites at Rusinga has yielded the fossilized fig with a bite mark on it—a mark of the right size to have been made by *Proconsul.* Short of finding a *Proconsul* fossilized with a fruit in its mouth, we couldn't ask for much better evidence of its diet.

Why would a frugivorous ape have a pattern of dental development so similar to—and as rapid as—that of a folivorous ape of similar body size (*Symphalangus,* the siamang)? Why wouldn't *Proconsul* have a slower dental developmental pattern more like that of the gibbon?

The answer that springs to my mind is: because *Proconsul,* the frugivorous ape in question, lived 18 million years ago. We can be reasonably sure that the ancestral condition in higher primates was to have a more rapid, more monkeylike dental development. Therefore, an ancient ape—a very early ape—should have more rapid dental development than a modern one, all other things, notably body size and diet, being equal. Logically, then, *P. heseloni* would have a more rapid dental development than the modern gibbon simply because *Proconsul* was still evolving the slowed pattern of development that is today typical of modern apes.

What is especially interesting is that the antiquity of *P. heseloni* gives it the same tempo of dental development as a modern ape (the siamang) with a folivorous diet. I guess there's more than one way to arrive at a given pattern of dental development.

There is a postscript to all these dental studies that is, in many ways, more satisfying to me than any of the other outcomes. The result has nothing to do with *Proconsul.* It has to do with one of those rare instances in which someone actually gets exactly what he deserves.

During their years of collaboration, Chris and David tended to get most of the credit because they were on the faculty of their respective universities while Don was unfairly considered "just a technician" because he held no advanced degrees. Then the University of Newcastle-upon-Tyne made a small but important change in its rules. Many British universities will permit a graduate of the institution to submit a bound set of publications, along with a 10,000-word essay on their significance, and be awarded a Ph.D. if a review of the dossier suggests the work merits such a degree. Where the University of Newcastle-upon-Tyne suddenly deviated from the norm was in no longer requiring that the applicant be a graduate of that university or, in fact, of any university. By the time this opportunity opened up, David had retired, but Don's head of department and dean encouraged him to compile a dossier and submit it. Don was duly awarded his Ph.D., which he richly deserves, and became a faculty member of the University of Newcastle-upon-Tyne. It was a sweet triumph of ability over educational opportunity.

11

More on Teeth

Having learned so much about *Proconsul heseloni* from these new techniques, I was eager to see what else might emerge from such studies. Would information about any of the other fossil primates help me refine my conclusions?

Unfortunately, despite the superb work carried out by Chris, David, and Don, we don't know M1 crown completion times for *Proconsul nyanzae* because they have studied only M2s so far. Second molars begin their mineralization process after birth, which means there is no neonatal line to show when birth occurred. From an isolated M2, there is no way to calibrate the histological data into days since birth. All that can be said is that the formation of the molar crowns in *P. nyanzae* took longer than crown formation in *P. heseloni,* which suggests that overall dental development was slower in the larger animal. Yet compared to the crown formation in a chimpanzee with roughly similar body size, that in *P. nyanzae* was incredibly fast, taking between 62 percent and 69 percent of the time needed to form those same teeth in chimpanzees. This fact points, once again, to the notion that these Miocene primates had evolved a slower pace of dental development than their immediate ancestors but that this Miocene pace was still faster than the modern tempo.

All of these observations raise a key question: How do we judge “living fast” or living slow” with regard to development and life-history strategy? Is there any way to decide if *Proconsul* grew like a monkey, like an ape, or like neither? I don’t think that in all cases the appropriate technique is to compare the absolute number of months needed to arrive at a particular developmental marker in two species, because we know that the absolute time of maturation is affected by body size, brain size, and, in at least some cases, diet. If we compare species that differ in any one of those parameters, we don’t know what to make of the results. And if we are studying dental development, then the absolute size of the teeth of each species is likely to affect the chronology as well. Instead, perhaps we should be comparing the percentage of total lifespan that is spent arriving at a particular stage of development. Unfortunately, determining the total average lifespan of an extinct species is very difficult.

In her work, Holly Smith developed another technique for comparing life-history strategies and dental maturation patterns in different species. She plotted out the information on the times of tooth crown and root formation of particular teeth in a living species, using a solid line to indicate crown formation and a dotted one to indicate root formation. Such a graph indicates, for example, how far an M2 has developed when the M1 of the same individual has completed its crown and has one-third of its root formed. On this graph, she then plotted information about the state of maturity of the teeth of a fossil individual (at the time of its death), marking the developmental stage of each tooth. If the pattern of growth is the same, then the points representing the fossil’s state of maturity will form a straight vertical line when they are superimposed on the living species’ data. If the pattern of growth is different—if, for example, there is less overlap in the development of adjacent teeth in the fossil species—then the points representing the fossil will make a crooked and/or slanting line. Mind you, this technique determines only similarity of *patterns* of growth, not absolute *chronology* or the number of months needed to form a particular tooth. Thus while, happily, this

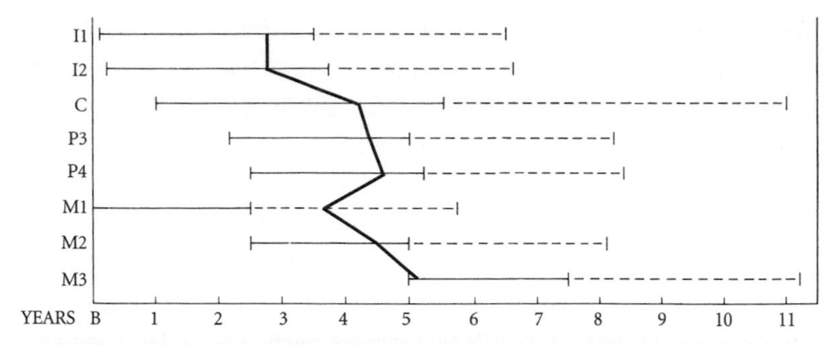

This diagram plots a single time in tooth development in *Proconsul* (dark zig-zagging line) against the standard developmental pattern in African apes. If *Proconsul* followed the same pattern, the dark line would be straight and vertical. The general left to right slant of the *Proconsul* line indicates that there was less time between the onset of development of one tooth and that of the subsequent tooth. Abbrieviations: I1 = first incisor; I2 = second incisor; C = canine; P3 = first premolar; P4 = second premolar; M1, M2, M3 = first, second, and third molars. Solid horizontal lines show the duration of crown formation; dashed lines indication the duration of root formation. B = birth.(Modifed from B. H. Smith, "The Physiological Age of KNM-WT 15000," in A. Walker and R. E. Leakey, ed., *The Nariokotome* Homo erectus *Skeleton* [Cambridge: Harvard University Press, 1993], 213.)

technique minimizes the effects of differences in body size, brain size, or tooth size, it cannot tell you how old the fossil was when its M1 crown erupted, for example.

Applying Holly's technique to *Proconsul heseloni* would let me evaluate its growth pattern. I used Holly's data on living species and the summary data from the British team on *P. heseloni.* I superimposed the development of crown and root formation for M1, M2, and M3 onto the chimpanzee diagram, temporarily ignoring the histological data that told me how many months had passed between birth and each particular crown completion. Although the *Proconsul* data did not make a perfectly straight and vertical line on the chimpanzee scale, they did not diverge far from this pattern. The diagram shows that the change that would transform the *Proconsul* pattern into a chimpanzee-like one is to increase the overlap between the development of the first and second molars. This sounds like a relatively straightforward genetic change in timing.

All signs point to the conclusion that *Proconsul heseloni* had a life-history strategy intermediate between those of living monkeys and living apes in a pattern that seems to closely resemble that seen in siamangs and gibbons. The data on *Proconsul nyanzae* are too patchy to permit any strong conclusions but this larger fossil ape may follow a pattern similar to that in *P. heseloni* at a slower rate of development. If an extended life-history strategy is indeed a characteristic and defining trait of the apes, *Proconsul heseloni* had already evolved a more apelike life-history strategy and there are hints that *P. nyzanzae* did too.

Another insight from these histological studies was the rate at which *Proconsul* deposited enamel (very rapidly) and dentine (very slowly) compared to such rates in other species. By measuring the distance between each adjacent pair of cross-striations on a particular tooth, David, Don, and Chris could calculate how much enamel was deposited in a day. They followed the same procedures to measure rates of enamel deposition in *P. heseloni* and *P. nyanzae* and in modern chimpanzees, gorillas, orangutans, gibbons, and gelada baboons. The results were surprising.

Chimpanzees, humans, and *Proconsul* showed similar patterns of enamel deposition. Early in the formation of a tooth, enamel is deposited fairly slowly, but as the cells making the tooth near the outer enamel of the tooth cusp, they start depositing enamel more rapidly. In *Proconsul,* the amount of enamel deposited each day rose from a little over 4 microns, as the tooth started to mineralize, and increased to about 7 microns a day near the surface. Compared to the other primates in the study, *Proconsul* had markedly faster enamel deposition. Even though *Proconsul* completed the entire process of making a tooth quickly, the enamel on its molars was relatively thick because its ameloblasts secreted a lot of enamel every day. You'd expect to see this effect most strikingly on the M1 because, among the permanent molars, it is the first to erupt and is therefore in wear for the longest period of time during an animal's lifetime. As you might predict, the enamel on the M1 of *Proconsul heseloni* can be classified as thick and that of *Proconsul nyanzae* is even hyper-thick. Hyper-thick enamel seems to be a sort of "nutcracker" adaptation that allows animals to crack the hard casings on nuts and fruits without breaking their teeth.

Though the enamel grew fast, the dentine underlying the enamel in *Proconsul* molars grew slowly, averaging only about 2.4 microns a day on the M2. This is slower than the dentine secretion rate in living great apes and humans, which average 4–6 microns per day. *Proconsul*'s dentine formation rate is close to that in the gibbon *Hylobates moloch,* which deposits 2.6 microns of cuspal dentine daily. This peculiar combination of rapid enamel deposition and slow dentine deposition is so far found only in *Proconsul.* But it explains perfectly how these animals made thick-enameled teeth rapidly.

Very few other Miocene hominoids have been looked at histologically. Jay Kelley, Tanya Smith (a graduate student at the Stony Brook University), Tanya's adviser, Lawrence Martin, and Meave Leakey of the National Museums of Kenya recently published papers on the dental development of one of my favorite species, *Afropithecus turkanensis.* The team estimated M1 eruption in *Afropithecus* because, unfortunately, the tooth could not be sectioned because it was still in place in a bony jaw and they didn't want to cut into the bone. At the time that the *Afropithecus* individual died, the M1 was not yet in wear. To solve this problem, they substituted data on M1 development and eruption from baboons, where these events could be observed directly, for the missing data on *Afropithecus.* In addition, two isolated M2s were sectioned.

They estimated that M1 emergence occurred in *Afropithecus* between 28.2 and 43.5 months. Consider the implications of these estimates. If M1 emerged in *Afropithecus* at the *minimum* estimated age of 28.2 months, this would be substantially later than M1 emergence in any species of monkey for which data are available. However, all living monkeys are smaller-bodied than *Afropithecus,* so the larger body size of the fossil might be responsible for a delayed M1 emergence. If M1 emerged at 43.5 months—an estimate that was based on an *average* value from living primates—then *Afropithecus* developed more slowly than an average modern chimpanzee, which would be most unexpected. The most probable true time of M1 emergence lies between these two values. In any case, both minimum and average estimates suggest that M1 emergence happened later in *Afropithecus* than in modern monkeys.

The amount of time needed to form the crown of M2 in the two *Afropithecus* teeth was calculated from direct counts of cross-striations. These were determined to be about 30 and 36 months, for an average of 33 months: about the same as a chimpanzee and some months later than *Proconsul nyanzae* (about 29 months) or *P. heseloni* (about 18 months). If all of the data and estimates on fossil dental development are reasonably accurate, then *Afropithecus* appears to show a life-history strategy that is slower and genuinely more similar to that in modern apes than in *Proconsul.* The data are still very limited though.

There are three known factors that might confound the issue of life-history strategy: antiquity, body size, and diet. Since *Afropithecus* and *Proconsul* are both 17–18 million years old, the slower life-history strategy of *Afropithecus* cannot be attributed to differences in antiquity. Because *Proconsul nyanzae* is about the same body size as *Afropithecus,* the latter's slower developmental tempo can't be explained away by body size differences. Finally, because *Afropithecus,* like *Proconsul,* seems to have been a nut and seed eater with thick enamel, diet can't be used to erase the difference in developmental tempo either. In short, the differences in dental development between *Afropithecus* and *Proconsul* appear to reflect an evolutionary difference in life-history strategy in and of itself. In this respect, *Afropithecus* seems to be a more modern Miocene ape than *Proconsul.*

An immense amount of time and exploratory work went into

developing these histological techniques for determining the age and life-history strategy of extinct Miocene primates—and the labor is beginning to pay off. There is a still a great amount to be learned and scores of reasonable questions cannot be answered as yet. But already we can say with some certainty that both *Proconsul heseloni* and *P. nyanzae* had begun to evolve an extended life-history strategy relative to modern monkeys, and that their pattern of dental development was similar to that of modern apes of the same body size. *Afropithecus* seems to have gone even further in lengthening its dental development and evolving a slowed-down life-history strategy. All three species show the sort of changes that we would expect to see in early hominoids as they evolved apelike lifestyles and apelike growth and development.

The comparison that helps sort out this confusion is between *Proconsul* and *Victoriapithecus,* which shows the differences and resemblances between a Miocene ape and a Miocene monkey that are roughly contemporaneous. Chris Dean and Meave Leakey have addressed this problem by sectioning two isolated second molars of *Victoriapithecus macinnesi* and adding information from eight canine teeth with visible perikymata.

Victorapithecus secreted enamel rapidly, at a rate of about 6 microns a day across its molar crown. *Proconsul* also secreted its enamel rapidly but showed a gradient in secretion rates going from slower secretion (4 microns a day) near the junction between the dentine and enamel to faster secretion (7 microns a day) in the cusps. Such a gradient occurs in living apes and monkeys but does not occur in two fossil lemurs that have been studied. Chris and Meave suggested that both the fast rate of enamel secretion and the lack of a gradient in enamel secretion might be primitive characteristics in *Victoriapithecus.* Another interesting finding was that, despite rapid enamel deposition, the fossil monkey *Victoriapithecus* took longer to grow its second molars and canines than a modern vervet monkey, which has a similar body size. Overall, Chris and Meave suggested that *Victoriapithecus* may match vervet monkeys in body size but not in dental development, in which it more closely resembles a macaque of somewhat larger body size. This evidence support the idea that *Victoriapithecus,* a species believed to be ancestral to Old World monkeys and apes, had a slower life-history strategy than would be expected for a monkey of its size, as Brenda Benefit has hypothesized. Possibly, then, throughout their evolution monkeys have speeded up their life-history strategies while apes have slowed them down.

Though the evidence is scattered and confusingly incomplete, my one-time postdoctoral fellow David Begun and others have tried to formulate a comprehensive hypothesis about the origins of modern apes. David identified *Afropithecus* as a plausible candidate for the ancestor of the adaptive radiation of hominoids in Europe in the later Miocene because it has thick tooth enamel and rounded, low-cusped molars. David suggests that these dental changes were a key adaptation that sparked the adaptive radiation of more modern apes, which would be able to exploit new and different types of food with such teeth.

This hypothesis is sometimes referred to as "In and Out of Africa." It is certainly intriguing. David's scenario has *Afropithecus* or something like it enlarging its geographic distribution about 16.5 million years ago, by moving out of Africa and into Europe as well. Either before or just after this geographic expansion—the fossil evidence is not yet good enough to settle the matter—the *Afropithecus*-like lineage began evolving into a more modern form. David calls this first expansion the *Griphopithecus* event, because the earliest known representative (17 to 16.5 million years old) of a more modern ape is a Miocene hominoid by that name from Engelweis, Germany. David believes that the thick-enameled hominoids with low-crowned teeth evolved, radiated, and spread in Eurasia for about half a million years. Then, according to his hypothesis, some of these more modern apes spread back into Africa between 16 and 15 million years ago, along with other mammals such as rodents, bovids (antelopes), giraffes, and carnivores. These second-time hominoid immigrants outcompeted and replaced the archaic African species that had stayed behind. Among these evolving "new" African apes lay the ancestors of the living African apes; probably the ancestors of the Asian apes were among those who stayed behind in Europe, though the fossil record is not clear. This idea of the connection from Africa through Arabia and on into Europe as a two-way street, busy with evolving species migrating in both directions, is stimulating.

One way to test David's hypothesis is to compare it to other indications of the timing of these events. Obviously, the fossil evidence provides the underpinning for David's scenario. But we always have to remember that we can judge only *from the fossils we have.* Where we do not have fossils—either in space or in time—we can draw only provisional conclusions.

Fossils of a particular type of animal (higher primates in this instance) may be missing for at least three reasons. First, perhaps the kind of creature we are interested in didn't live in that place at that time. Second—and this is an important alternative—fossils may be absent because fossilization did not occur at the place and time where the animals lived. Geographic accident has certainly robbed us of part of the potential fossil record many times; only a tiny percentage of all of the animals that ever lived have been turned into fossils, and fewer still are found. Finally, if fossilization occurred, then the fossiliferous beds which preserve the evidence from that place and time might be buried so deeply in the earth (or might be in an area that has not yet been explored for fossils) that we know nothing of them yet.

How else can we test David's imaginative and intriguing hypothesis?

Where the fossil record is conspicuously silent or confusing, we can turn to molecular evidence. In 2001, the results of a collaborative study were published by Rebecca Stauffer, a student at Pennsylvania State University, with her adviser, Blair Hedges, who is a renowned molecular biologist at Penn State, his technician, Maureen Lyons-Weiler, and me. We used specimens from Oliver Ryder of the San

Diego Zoo. The basic premise of the project is a common one: the longer two species have been separated evolutionarily, the more genetic changes or mutations will separate them. Therefore, by counting the number of molecular differences between two species, you can estimate the date at which their lineages diverged from each other. If mutations occur at an approximately constant rate, then—like the ticking of a clock—each mutation represents a set unit of time.

Constructing an accurate molecular clock requires a lot of technical skill in the laboratory and tremendous care in data collection, molecular sequencing, and statistical analysis. It also requires a good knowledge of the fossil record because every molecular clock must have a calibration point—a divergence of known date—and that can only be derived from the fossil record. In our study, we used an earlier molecular clock dating of the divergence of Old World monkeys and hominoids at 23.3 million years ago as the calibration point. This point also happens to be close to the end of the Oligocene epoch and the beginning of the Miocene.

The boundaries between geological epochs were often times of major upheaval or catastrophic climate change. Of course, geological epochs are a human construct. Logically, their boundaries were set at points in the stratigraphic sequence when there were significant, visible, and widespread changes caused by those climatic or geological events. Another consequence of these upheavals is that an unusually high number of extinction and speciation events coincide with epochal boundaries. It is not surprising that the fossil record indicates that the divergence between Old World monkeys and apes occurred about 23 million years ago, one of those epochal boundaries. The first Miocene ape fossil—the first tangible evidence that the ape lineage had diverged from the monkey lineage—is dated to 21 million years ago.

As a check, we repeated our calculations using two other calibration points. One was the divergence between primates and rodents, about 110 million years ago, and the other was the divergence between carnivores (animals such as lions, bears, and hyenas) and artiodactyls (herbivores with an even number of toes, such as deer, buffalo, or pigs) 92 million years ago. The idea was that if one of our calibration dates was wildly wrong, the results from that set of calculations would differ markedly from the others and tip us off to the problem.

Blair and our collaborators in his lab sequenced DNA from nine nuclear genes in a gorilla and in an orangutan and added these results to already-available data on the same genes in humans, common chimpanzees, and gibbons. They also analyzed the complete sequence of the mitochondrial DNA from humans, common chimpanzees, bonobos, gorillas, gibbons, and one monkey, a hamadryas baboon.

Understanding the difference between mitochondrial and nuclear DNA is important. Nuclear DNA is the one that was first discovered; we all learned about DNA and its classic double-helix structure in our first biology class. Nuclear DNA makes up the chromosomes that pair up, replicate, and reassort during sexual reproduction. Thus an offspring will get half of its nuclear DNA from its father, through his sperm, and half from its mother, through her egg—which is why none of us is exactly like either of our parents.

Mitochondrial DNA, or mtDNA, works differently. Mitochondria are organelles that live in the cytoplasm of the cell, not in the nucleus. They are often said to be the power stations of the cell, the source of energy that the cell uses to grow and divide. But the DNA in mitochondria does not determine what the possessor of those genes will be like (unlike nuclear DNA), and mtDNA is passed from parent to offspring by a different mechanism. During sexual reproduction, the father's sperm fuses with the mother's egg, so each parent contributes half of the nuclear DNA. But the egg contributes lots of mitochondria to the new offspring, whereas the sperm contributes few or no mitochondria. Simply put, all of the mtDNA that is passed to the offspring comes from the mother, and from her mother before her, and from her mother before her. Generation after generation, a mother passes her mtDNA to both her sons and her daughters, but only her daughters can in turn pass that same mtDNA on to their daughters, and so on. Thus mtDNA can be thought of as a message that encodes a personal, female lineage back through time. Like nuclear DNA, mtDNA also evolves and mutates through the generations—and like changes in the nuclear DNA in genes, the mtDNA can be used as a molecular clock to date evolutionary divergences.

Why use two DNA clocks? Because they are subject to different sorts of errors and influences.

With nuclear genes, the number of possible mutations is limited by functional demands. For example, if a mutation in a gene means that the correct enzyme is no longer produced, the organism dies and its mutated genes are not passed on to future generations. By sequencing many genes (the multigene method) and averaging the dates they yield, we are likely to arrive at a more accurate divergence date than if we use only a single gene.

If DNA is to work as a clock, then mutations have to occur with clocklike regularity—and the assumption that mutations occur in mtDNA at a constant rate has been challenged. There is significant evidence that in primate lineages the mutation rate of mtDNA has speeded up over time. Divergence dates calculated from mtDNA thus may be older than they should be. By using both nuclear DNA and mtDNA, we were able to compare the answers from these two different sources and hope to get close to the truth. We know that dates from molecular clocks usually have large errors and we have to be careful.

From the genetic data we derived a branching pattern and estimated some important divergences in the primate fossil record. No matter which dataset or which calibration method we used, we always got the same branching sequence and we got extremely similar dates for the divergences, too. The first divergence from the common hominoid stem was the lineage of the gibbons and siamangs, about 14.9 million years ago. Next, the orangutan diverged from the com-

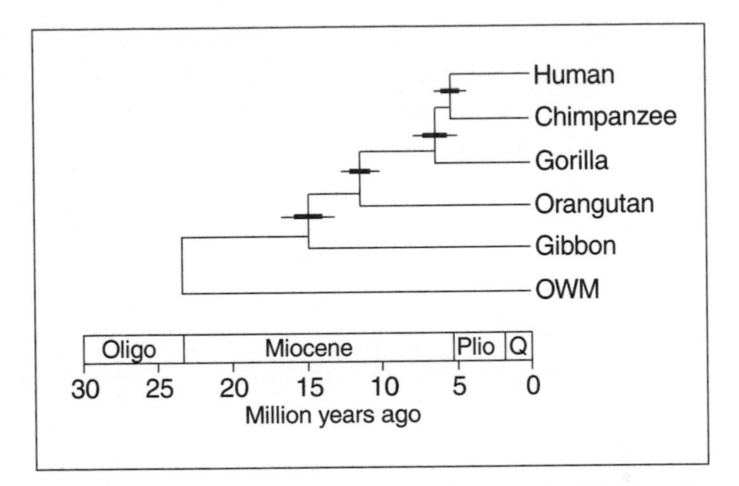

This diagram shows the estimated times at which various ape lineages split from the stem monkey-ape lineage, on the basis of molecular differences. Time estimates are shown with ± 1 standard error (heavy bars) and 95 percent confidence interval (narrow bars). Abbreviations: Oligo = Oligocene; Plio = Pliocene; Q = Quaternary. (From R. L. Stauffer, A. Walker, O. Ryder, M. Lyons-Weiler, and S. B. Hedges, "Human and Ape Molecular Clocks and Constraints on Paleontological Hypotheses," *Journal of Heredity* 92, no. 6 [2001]: 471; reproduced by permission of Oxford University Press.)

mon stem about 11.3 million years ago. The third divergence was that of the gorilla lineage, about 6.4 million years ago. The fourth and final divergence occurred between the chimpanzee lineage and the human lineage, and occurred at about 5.4 million years ago. None of these divergence dates should be taken as precise; they simply reflect the average data within a range of possible dates. Compared to radiometric dating of rocks, genetically derived dates have large ranges of error.

The point of all this genetic work and calculation is to be able to put some constraints on the fossil record, especially in cases where the interpretation of the fossils is controversial. Genetics will never give us specific information about the morphology of an extinct species—not until we can extract DNA from fossils readily and know a great deal more about which genes do what. But we can use the genetically derived branching sequence and divergence dates to eliminate impossibilities.

Going back to David's hypothesis, we can see if the timing of the events he proposes matches up with the molecularly derived dates. I will summarize the arguments briefly. David suggests that *Afropithecus* or another early Miocene ape may have been the hominoid species that migrated out of Africa into Europe about 16.5 million years ago and then gave rise to more modern-looking hominoids. The features that make *Afropithecus* a good candidate are its antiquity (it lived 17–18 million years ago), its enamel thickness (about the same as in *Proconsul* and thicker than the enamel in living great apes), and its slowed life-history strategy. Because *Afropithecus* has such an extremely long and unusual face, David thinks it may be too derived or specialized to have been ancestral to the later, European forms.

If the molecular dates are correct, then the migration of the more modern hominoids from Europe back into Africa occurred just before the divergence of the gibbon/siamang lineage. Did the early gibbons stay behind in Eurasia or did they evolve in Africa and migrate back out for a second time? When pressed on this point, David smiles broadly and confesses that he can't decide without more fossil evidence.

There are several African hominoids dated to about 15 million years ago and the best-known species, *Nacholapithecus kerioi* from Samburu and *Equatorius africanus* from Kipsaramon in Kenya, are represented by partial skeletons. *Nacholapithecus* had an interesting combination of characteristics. It was tailless—an apelike feature—and had a long and flexible vertebral column, more like a modern monkey's. Its feet were long-toed, for grasping, and its forearms were long: Both these features are often considered indicators of a brachiating habit though nothing suggests that *Nacholapithecus* had the extreme and pronounced adaptations of a true brachiator such as a gibbon. *Equatorius,* though of similar age, has postcrania that tell a different story. It, too, has a long, flexible vertebral column but its limbs are adapted for quadrupedal locomotion on the ground. Neither *Nacholapithecus* nor *Equatorius* looks like a gibbon ancestor or anything close to it. In Asia, the fossil record is even scarcer and is similarly silent on the early ancestry of gibbons and siamangs.

The next divergence involves the splitting off of the orangutan lineage about 11 million years ago. There are several potential ancestors for modern orangutans, or at least a number of fossils that must be closely related to orangutans.

A collection of fossil teeth from Thailand, recently announced in the journal *Nature*, represents a previously unknown species called *Lufengpithecus chiangmuanensis. Lufengpithecus* comes from a geologic bed sandwiched between two dated layers. The upper layer is 10 million years old and the lower is 13.5 million years old, so the fossils must be older than 10 million years and younger than 13.5 million years. If the true date of the fossils lies closer to 10 million years ago, then new *Lufengpithecus* might have been ancestral to modern orangutans, as the Thai-led team claims. Unfortunately, so far there are only teeth from Thailand. Other specimens and species of *Lufengpithecus* between 9 to 8 million years old are known from China. The teeth of these Chinese specimens look very similar to the teeth of the Thai specimens and modern orangutans, but the skulls of the Chinese *Lufengpithecus* are disappointingly unlike orangutan skulls.

The most likely orangutan ancestor yet found is another new specimen from Thailand, a fossil jaw called *Khoratpithecus* dated to between 9 and 7 million years old. Once again, though, more complete fossils are needed before we can be certain.

Until this new Thai jaw was found, the most likely orangutan ancestor was *Sivapithecus* from Pakistan and India, dated to 12.8–8 million years old. When *Sivapithecus* jaws, teeth, and faces were first discovered in the 1980s, they seemed so similar to those of modern orangutans that *Sivapithecus* was widely hailed as the ancestor of the orangutan. But, as the number of fossils increased, more differences began to appear. One of the most problematic is the fact that *Sivapithecus* limb bones show it walked quadrupedally on the ground rather than hanging, clambering along, and slowly swinging from branches the way orangutans do today.

The estimated divergence dates lead us to hope we'd see gorillalike fossils at about 6.5 million years ago and the first chimpanzee-like hominoids at about 5.5 million years ago. Unfortunately, we have virtually *no* fossils of any age that tell us anything useful about gorilla or chimpanzee lineages.

All of this means that we can neither prove nor disprove David's In and Out of Africa hypothesis. We certainly don't find fossils that look like the early ancestors of gibbons, gorillas, or chimpanzees at the right place and the right time. And the Miocene fossils of orangutan-like apes are either the wrong age or certainly don't possess the postcranial adaptations that would solidly demonstrate they were ancestors of modern orangutans. One of the questions we cannot yet answer is whether the highly evolved postcranial adaptations of gibbons, siamangs, or orangutans were there from the beginning of those lineages.

The molecular data have been useful in clarifying another ambiguous case involving *Morotopithecus,* a large-bodied Miocene ape from Uganda found in sediments dated to 21 million years ago. *Morotopithecus* is the new name given to the large ape vertebrae from Uganda that Mike Rose and I wrote about and Carol Ward later studied as *Proconsul major.* One suggested interpretation of this fossil was that it was an early great ape ancestral to gorillas, chimpanzees, and orangutans but not to gibbons or siamangs. The alternative interpretation is that *Morotopithecus* was a stem hominoid ancestral to all modern apes including gibbons and siamangs. Since the molecular evidence puts the divergence of the gibbon/siamang lineage at about 5 to 6 million years after *Morotopithecus* lived, the "stem great ape" hypothesis looks much weaker and the "stem hominoid" hypothesis is strengthened. Although molecular dates are calculated rather than measured, to be wrong by 6 million years would be surprising.

These findings underscore the disadvantage of looking backward from the present toward the past. When we look at living great and lesser apes, we perceive a total package involving a complex combination of locomotion, anatomy, ecology, and life-history traits. The natural tendency is to take that ape package as a standard or norm. When we peer dimly into the past history of apes, we simply cannot see a single abrupt change that transforms some sort of crude proto-ape into a modern one and we are dissatisfied or confused. What we see instead is a jumble of developments and adaptations that occur at different times. They seem to be out of sync to our eyes, since we are unconsciously seeking an ancient yet fully modern ape. The disjunction between our expectations and our observations is a warning, a caution to think carefully before we evaluate the past. Modern species have had much longer to evolve than ancient ones, for the simple reason that the ancient ones died long ago and the modern ones are alive or dead *now.* All the evidence suggests the complex of traits we see in species today did not evolve all at once in the long-distant past but came about in a piecemeal or mosaic fashion at different times.

We should have expected that. But glimpsing the past always provides a bit of a surprise, even when you can rightly say you expected something like that. And it is those surprises that provide the fun of paleontology.

12

Listening to the Past

When I look back at my career-long association with *Proconsul,* I am surprised and amused by the varied twists and turns it has taken. Teeth and jaws have played an important role in our growing knowledge of these ancient hominoids, from the size and proportions of teeth right down to the tiniest microscopic detail of how the teeth were formed during development. But teeth sometimes mislead us; teeth and jaws are not the only important features of a species, past or present.

Because we had the tremendous luck to find partial skeletons of *Proconsul heseloni* and *Proconsul nyanzae,* my colleagues and I were given unusual opportunities to try to reconstruct and understand the way these Miocene apes moved about their world. Locomotor studies on living and fossil primates were one of my first obsessions, one I "caught" from John Napier. I like to work back and forth between the fossils and our ever increasing knowledge of the locomotion of the living species, for the more I know about the locomotion, posture, and habitual movements of living animals, the better able I am to interpret the past.

Even the most knowledgeable functional anatomist cannot help wondering about errors and misinterpretations, though. Observing anatomical details, quantifying them, and understanding what difference they might make to the animal involved is a fairly straightforward matter of training and analysis, improved by the addition of as much intelligence and creativity as can be mustered by the analyst. Yet there is always a question of how much emphasis to give to any particular feature, or group of features: witness the changing assessments of the adaptations for brachiation in the *Proconsul*-in-a-tree. John Napier and Peter Davis had almost a complete forelimb to work with. They emphasized the adaptations to brachiation they saw in that forelimb, as did a number of researchers who reanalyzed that material after them. By the time we had recovered almost the complete skeleton, the emphasis on brachiation and suspensory locomotion—hanging by the arms from the trees—had given way almost completely to an interpretation that the same specimen was a slow, clambering creature that most of the time walked quadrupedally on top of branches.

John was inventing the field of locomotor analysis in 1959 when he and Peter worked on the *Proconsul*-in-a-tree. It is all very well to say that our techniques and analytical methods have improved vastly since then and to use that as an explanation for why we are right now and they were not. But every time I find myself making an excuse like that, I wonder where the field will be in 50 years' time and what improved methods we'll have by then. I always hope to find an independent way to check our conclusions that might point up mistakes in our reasoning.

If an ecological reconstruction can be made of a fossil site with the aid of geological data, plant and insect remains, and the inferred habits of nonprimate species, this can serve as a sort of test of the locomotor hypothesis that has been developed from the primate fossils. With the superb and abundant preservation of remains that we have at Rusinga and Mfangano, from the tree trunks themselves to fruits, vines, nuts, leaves, insects, and many different species of animal, we don't have to question our interpretation of *Proconsul* as a heavily arboreal species. At the very least, the exceptional preservation of all these sorts of remains suggests strongly that *Proconsul* used forest resources and lived and died in wooded areas.

But could we find another truly independent means of determining the locomotor habits of extinct species? Thanks to my old friend Fred Spoor, I have. Fred is a tall, lanky Dutch anatomist with a shaved head, in the fashion of many modern athletes.

One of the first times Pat and I met Fred was in 1988 at a conference in Sardinia, which was exceptional not only for the interesting papers that were delivered but also for the abundant, even overwhelming, quantities of food and drink that were served to the conference participants at every meal. The sun was hot, the surroundings were gorgeous, the company was good, and the refreshments were delicious. Frankly, it was not unusual to see one or more of the participants staggering off for a nap after lunch, not to reappear until hours later. On one memorable day, the beautiful Mediterranean sunshine disappeared and a thundering rainstorm moved in—provoking some of the participants from New Zealand to perform a traditional Maori *haka.* The *haka* is a dance normally used by the Maori in preparation for battle or warfare; it is now performed by the New Zealand rugby team, the All Blacks, before every match as well. The *haka* is loud, ferocious, and intimidating, as a war dance should be. In Sardinia, it seemed to have a positive effect on the weather. Watching this spontaneous demonstration, and watching the bemused Sardinian hotel staff watch us watching the dancers, was marvelous. I will never forget it.

Fred now works at University College, London, together with Chris Dean. Whereas Chris has focused on the detailed anatomy of teeth and what they can tell us about past and present species, Fred has become an expert in other areas. Perhaps his favorite two topics are various sorts of imaging techniques and the anatomy of the skull as it can be studied in great detail through imaging. In the course of completing the research for his Ph.D. from the University of Utrecht in 1993, Fred learned how to use CT scans—what the layperson calls CAT (computed axial tomography) scans—to look inside the bony ears of fossils. CT scanners take serial images or virtual slices through an object, in this case, a fossil skull. These slices can be reassembled into a three-dimensional reconstruction of the original, with the help of a complex computer program. The three-dimensional image or any part of it can be measured to give highly detailed information about structures that cannot be seen on the surface.

From some rather obscure anatomical studies in the nineteenth and early twentieth centuries, Fred realized that the bony inner ear records in its shape important information about how an animal moves and stands. The earlier anatomists had either dissected skulls or injected the bony inner ears with wax or latex and then dissolved the skulls away to take their measurements. These techniques would not normally be available to someone working with fossils because one of the most important charges of curators is to prevent damage to specimens. But, Fred realized, the advances in imaging techniques in the latter part of the twentieth century made it possible for him to make comparable measurements without cutting or otherwise damaging a skull.

Fred took a basic anatomical principle—that the inner ear is functionally related to how an animal moves—as the rationale for studying a series of fossil skulls to gain a new type of information about ancient locomotion. At first Fred's work involved persuading curators to let him take their precious specimens out of the museum and then coaxing clinical radiologists at hospitals into resetting their CT scanners to make images of the fossils. Fred's eagerness, intelligence, and charm stood him in good stead as he learned how to take, store, manipulate, and interpret the images.

In recent years, Fred and I have been working with micro-CT, using a special machine at Penn State, where I work, that takes virtual slices of rock or fossil or bone as thin as 1/5000th of an inch (about 1/100th of a millimeter). The software has been elaborated and refined, so that the three-dimensional model can be rotated on the computer screen and very accurate measurements can be taken.

But what does the ear have to do with locomotion? Well might you ask. The inner ear has to do with both hearing and balance, or equilibrium. The anatomical structures involved in hearing are the outer ear—the fleshy flap on either side of your head—the middle ear, a cavity in which lie tiny bones (the ear ossicles) that sense and amplify vibrations, and the inner ear, which houses both part of the hearing apparatus and the organs of balance, or equilibrium.

The way the organ of balance operates is elegant. As an animal moves through space—up, down, sideways, at fast, slow, or intermediate speeds—the animal needs to sense where it is and how fast it is moving. Animals also need to be able to keep the world in focus as they move. For example, when an animal leaps from branch to branch high above the forest floor, being able to see and estimate the three-dimensional location of the landing branch precisely is of tremendous importance. If high-speed motion makes your vision blur, a serious fall may be in your immediate future. The mechanisms that lets an animal's eyes record the moving world with as little displacement as possible—which let an animal sense accurately where it is relative to unmoving objects—are called the vestibular reflex. Even in slow-moving animals, this reflex is crucial for survival.

A simple exercise will demonstrate the operation of the visual reflex. Hold your head still, then take this book and move it rapidly from side to side as you try to read this sentence. Can you still read the words? No. Now keep the book still and move your head rapidly from side to side. Magically, your ability to read is not impaired even though your head is moving as fast as the book moved. Why? Because there is a mechanism in your inner ear that informs your brain how and how fast your head is moving, so that the eye and neck reflexes can keep the gaze steady and the visual input coming to your brain can be interpreted correctly.

How does this mechanism work? Inside the petrous bone, on each side, lies a structure known as the bony labyrinth. Toward the front of the bony labyrinth is a spiral structure that resembles a snail shell; this is the cochlea, which is part of the hearing apparatus. Above it are three canals—the anterior, posterior, and lateral semicircular canals—that loop through the petrous bone like tiny tunnels. These three canals are joined together and, in life, house a tubular mem-

Data from a micro-CT of a dried skull of a bushbaby, reconstructed by Fred Spoor, show the anatomical details of the bony labyrinth. The three semicircular canals are the looping structures at top right. The spiral cochlea, shaped like a snail, is at the bottom left. Animals such as bushbabies that are agile and move rapidly have larger semicircular canals. (Image by Fred Spoor using Voxel-man; © Fred Spoor.)

brane that is filled with a fluid called endolymph. When the head and body move, the canals move but the endolymph tends to stay still because of inertia. This discrepancy in movement is picked up by a particular region of each semicircular canal called the ampulla, in which the membrane's inner surface is studded with tiny hair cells that are surrounded by gelatin. Moving the head causes the endolymph to push against the gelatin, bending the tiny hairs and activating a nerve at the base of each hair. Those nerves then signal the brain that the body is moving, in which direction(s) it is moving, and how fast it is moving. Thus the hair cells act as miniature motion detectors that inform the brain how to compensate for the movements. This information keeps the visual systems working properly and the body's sense of balance intact. Because the three canals are oriented in three planes that are approximately orthogonal (at right angles) to one another, movement in any plane can be detected. It is an elegant and ingenious system.

From painstaking anatomical studies of many species, we know that specific features of the semicircular canals encode information about locomotor habits. Even though the membrane and fluid decay and disappear after death, the bony canals themselves are preserved in dry skulls or in fossils. One of the most important features is the size of the arc described by each canal, or its radius of curvature. Animals that move faster and more agilely have semicircular canals with larger radii of curvature—each canal approximates a bigger circle. Slower, more deliberate species have smaller radii of curvature of their semicircular canals. Thus gibbons and siamangs that brachiate rapidly from branch to branch have canals with larger arcs than those in the great apes, which move more slowly.

As with so many anatomical features, though, you have to consider absolute body size if you want to compare two species. There is a small but significant increase in arc size as body size increases that might be mistaken as the effect of more rapid movement.

Another caution is that, as yet, there is no simple way of quantifying the locomotion that an animal habitually uses or used from measuring its semicircular canals. We have to use fairly simple and gross categories, such as "agile, maneuverable, and acrobatic" versus "slow-moving and cautious." Even these provide enough information to serve as independent evidence of locomotor habits in fossils, though. For example, the three-toed sloth, *Bradypus tridactylus,* is one of the slowest-moving and most deliberate animals. Sloths live in the trees and hang suspended below tree branches for hours on end, without moving, which is probably why they were named after the proverbial sin of laziness and inactivity. Sloths move so slowly and so infrequently that they actually grow algae on their fur, like a tree trunk. Once, in the tropical rain forest habitat at the National Aquarium in Baltimore, I saw a sloth moving at what must have been its top speed and, even then, the creature seemed to be moving in slow motion. As you might expect, a sloth has semicircular canals that have an exceptionally small radius of curvature.

At the other extreme is the tarsier, *Tarsius bancanus,* a small nocturnal prosimian that lives in the forests of southeast Asia. This tarsier moves rapidly, leaping acrobatically from branch to branch at remarkable speed, and compensates for this motion with quick head movements. The compensatory head movements seem exaggerated in tarsiers because its eyes, like those of owls, are almost immobile within the bony orbits. During locomotion and even when tarsiers are at rest, looking for insect prey, their head movements are very rapid and noticeable. As expected, tarsiers have very large semicircular canals with large radii of curvature.

When we began our collaboration, Fred already had a small comparative dataset that he had collected for his thesis work. I wanted to know how the semicircular canals of *Proconsul* compared to those of living species. First we tried to micro-CT a petrous bone from the Kaswanga Primate Site at a commercial facility in Austin, Texas. The results were unhelpful because it was too difficult to distinguish the fossil bone from the limestone that had filled the canals.

After securing Meave Leakey's permission as Head of Palaeontology at the National Museums of Kenya, I returned to Nairobi to carefully drill away the bone from around the limestone-filled canals on this specimen. I used an airscribe and a binocular microscope so I could work precisely. At every stage of this operation, I stopped to take detailed photographs and to make high-quality replicas of the surface in case anyone ever needed to know exactly where the bone had been. Finally we were able to measure the radii of curvature of the canals.

I didn't expect that the locomotion of *Proconsul* was going to match the extremes in Fred's comparative sample, but I wanted to carry out an independent test of the anatomical and biomechanical studies my colleagues and I had done on *Proconsul,* by examining its semicircular canals. The hypothesis we had developed from our previous work on *Proconsul* limbs and skeletons suggested it was a slow-moving, deliberate, arboreal species roughly similar in locomotion to the living howler monkey of South America. The earlier work by John Napier and Peter Davis, among others, had emphasized an alternative view: that *Proconsul* had been arm-swinging or brachiating in the trees to some extent. Which would prove correct?

Satisfyingly, the radii of curvature of the semicircular canals of *Proconsul* fall neatly into the middle of the measurements on modern, slow-moving monkeys, not apes. *Proconsul* clearly didn't have

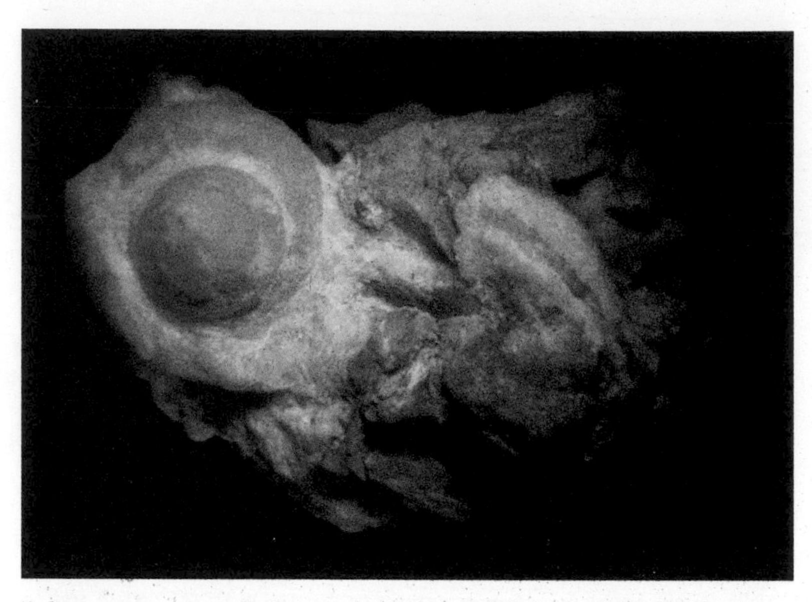

Delicate preparation of a *Proconsul* specimen revealed the calcite-filled cochlea at the left and one semicircular canal at the right. (© Alan Walker.)

large semicircular canals typical of agile, swift-moving modern apes such as gibbons or siamangs. If *Proconsul* did any swinging through the trees by its arms, it didn't do much and it probably wasn't very good at it.

This finding confirmed our earlier conclusions, which is always pleasing, but the results also reemphasized another much bigger point. We had discovered that stem apes such as *Proconsul* weren't like modern apes; they were more like some modern monkeys, but the full truth was that they were like no group alive today. All the great people who contributed to the first publications on *Proconsul*—John Napier, Louis Leakey, Le Gros Clark, Peter Davis, and Arthur Hopwood—had been trying to answer the question: *Was Proconsul an ape or a monkey?*

Now we knew that wasn't even the right question. We didn't need to know whether *Proconsul* was an ape or a monkey, as defined by

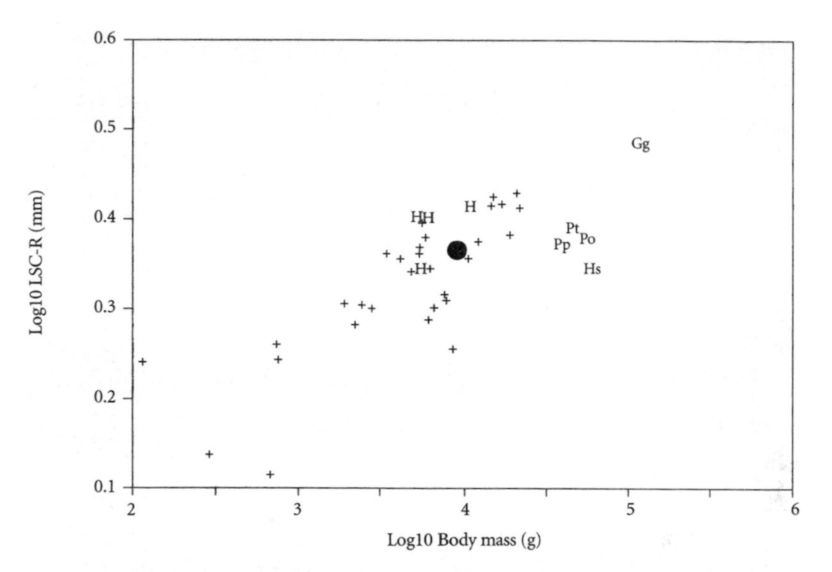

The size of the radius of the lateral semicircular canal (LSC-R) is correlated with the body mass of each primate species. Species indicated by + are monkeys or prosimians. H = gibbon; Pt = common chimpanzee; Pp = bonobo; Po = orangutan; Gg = gorilla; Hs = human. *Proconsul* is indicated by a large dot that falls in a cluster of quadrupedal Old World monkeys. (© Alan Walker.)

modern species, because it wasn't possible for apes in the Miocene to be like apes today. If you think carefully about the process of evolution, you know Miocene hominoids *couldn't* have been like modern apes unless evolution had stopped at the end of the Miocene.

What we need to focus on was what *Proconsul* and other stem apes were like in their own right, as early, ancestral apes. Then we could use that information to redefine the diagnostic attributes of a creature in the category "ape." In an important way, studying *Proconsul* reshaped our whole notion of how we study evolution and what we want to know about the past. This work has begun to free us from the pull of the present. We can now try to establish a perspective from which we can see the past for what it truly was.

A century ago, the chimpanzee Consul performed on the stages of Paris and London and caused his audience to rethink what an ape was and how it differed from a human. He bore haunting resemblances to humans in his behavior and dress, yet he was, eternally and fundamentally, an ape. It was a distinction then thought to be vast that now seems much more trivial. We know from genetic studies that we share with chimpanzees an overwhelming proportion of our genes, perhaps 99 percent. There is such an incredible underpinning of common biology, behavior, anatomy, and ancestry that links us to apes that sometimes we do not seem to differ from them by much at all.

Now Consul's fossil namesake, *Proconsul,* has pulled the same trick on us. *Proconsul* has redefined "ape" and redefined "monkey." Many of the intricacies of *Proconsul's* growth and development, brain size, locomotion, diet, and ecology have been preserved and can be read in the fossil record, when we are lucky enough and clever enough to develop the right techniques. By revealing its secrets to us, *Proconsul* has shown us much more than we ever expected to know about how the common ape-human lineage began and evolved.

To be human is also to be an ape. And in a very real way, to be an ape is to be *Proconsul.* I count myself fortunate indeed to have been involved with this ape in our family tree.

Epilogue

We know now that *Proconsul* in its various species anchored the hominoid lineage between 21 and 14 million years ago, revealing our beginning as primitive and in some ways monkeylike creatures. However much we have learned about *Proconsul,* it is still not nearly as much as we know about the modern end of the hominoid lineage: humans and the living apes, greater (gorillas, orangutans, and chimpanzees) and lesser (gibbons, siamangs). Between hominoid beginnings in the early Miocene and the present, a great deal of evolution occurred that is as yet poorly known.

As this book was readied for press, an exciting partial skeleton of a Miocene ape was unearthed in Spain that offers a stunning glimpse of the empty middle period of hominoid evolution. *Pierolapithecus catalaunicus*—named for the Catalonia region and the site itself, Els Hostalets de Pierola—lived between 13 and 12.5 million years ago.

Known from a single, adult male specimen, *Pierolapithecus* in life weighed about 66 pounds (30 kilograms). With its large, almost banana-shaped canines and long, sloping facial profile, the skull and teeth of *Pierolapithecus* are reminiscent of *Afropithecus* or even *Proconsul.* Some details of the new ape's facial anatomy, however, are less primitive and link it most closely to the living great apes. The discoverers believe that *Pierolapithecus* evolved only after the lesser apes (gibbons and siamangs) diverged from the common hominoid stem, an interpretation supported by the fossil's age.

In another important resemblance to great apes, *Pierolapithecus* is the oldest fossil hominoid with a broad, shallow torso and short, stiff lower back. The curvature of its ribs shows the shape of the trunk and its long clavicle indicates that the shoulder blades were set on the back of the rib cage. The vertebrae in the lower back of *Pierolapithecus* had very limited mobility. The features of the trunk in *Pierolapithecus* are functionally linked to the great shoulder mobility of modern apes, which is used both in brachiation and in sitting in an upright posture and reaching out with the hands. Like modern brachiating apes, *Pierolapithecus* had a wrist joint adapted to mobility, not stability. In all of these features, *Pierolapithecus* differs sharply from the primitive condition seen in *Proconsul,* with its deep torso, long flexible back, and more stable wrist.

If we had no additional limb fossils, we might predict that *Pierolapithecus* was fully adapted to brachiation—and we would be wrong. The hands of *Pierolapithecus* tell a different story. Whereas brachiating apes have elongated hooklike fingers made up of strongly curved phalanges, *Pierolapithecus* has relatively short and straight fingers, clearly adapted for palm-down walking like a monkey.

How can we reconcile these conflicting points of anatomy? Clearly *Pierolapithecus* was adapted for an upright sitting posture, with highly mobile arms, and yet it was not adapted for brachiation. The unexpected mosaic of characteristics in *Pierolapithecus* demonstrates that what we once thought of as a single package—"adaptation for brachiation from trunk to fingertips"—evolved piecemeal. Yet again, we have been fooled by the pull of the present.

We must hope for more spectacular finds and more surprises.

Pronunciation of African Words and Place Names

Allia Bay	AHL-ee-yah BAY
Buluk	bull-OOK
Bunyoro	bun-YORE-oh
Entebbe	en-TEB-bee
Gumba	GUM-bah
Hiwegi	hih-WAY-gee
Homa Bay	HOME-ah BAY
Kalodirr	KAL-oh-deerr
Kamasengere	KAH-ma-sen-GEER-ee
Kampala	kahm-PAH-la
Kanam	kan-AM
Kanjera	kan-JEER-ah
Kariandusi	KARE-ee-an-DOO-see
Karungu	kah-RUNG-goo
Kaswanga	kas-WANG-ah
Kavirondo	KAH-vih-RON-do, with a short a in "kah"
Kiahera	KEE-ah-HER-ah
Kiakanga	KEE-ah-KANG-gah
Kipsaramon	kip-SAR-a-mon
Kisingiri	KISS-in-GEER-ee
Kiwegi	kih-WAY-gee

Koobi Fora	KOO-bee FOR-ah
Koru	KORE-oo
Legetet	LEG-uh-tet
Maboko	mah-BOW-ko
Machakos	mah-CHA-kos
Mbita	um-BEE-tah
Meswa Bridge	MESS-wah BRIDJ
Mfangano	muff-ahn-GAH-no, with a very short u in "muff"
Moroto	more-OH-to
Nairobi	nye-ROW-bee
Napak	nuh-PAK
Ndege	en-DAY-gee, with a hard g
Nditi	en-DEE-tee
ngule	en-GOO-li
Nyanza	nye-AN-zah
Ol Donyo Lengai	ohl DON-yo LEN-guy
Olduvai	OLD-oo-vai
Oldoway	OLD-oh-way (old spelling of Olduvai)
Olorgesailie	OL-or-geh-SAL-lee
panga	PANG-ga
Rusinga	roo-SING-ah
Samburu	sam-BOO-roo
Sena	SAY-na
Siboloi	SIB-oh-loy
Songhor	SONG-orr
Tanganyika	TAN-gan-YEE-ka (colonial name of Tanzania)
Tinderet	tin-der-ETTE
Turkana	tur-KAH-nah
Ukamba	oo-KAM-ba

Notes

Archives KNM *Archives of the Kenya National Museum, Nairobi*
Archives NHM *Archives of The Natural History Museum, London, formerly the British Museum (Natural History)*

Prologue

Page

1 "He caused a sensation": Anonymous ("L. R."), "Le chimpanzee Consul," *La Nature: Revue des Sciences et de Leurs Applications Aux Arts et à l'Industrie* 1591 (1903): 415–516.

3 "a poem, written for the occasion": Ben Brierly, "In Memory of 'Consul,' the Belle Vue Chimpanzee Who Died Nov. 24th, 1894, Aged about 5 years," broadsheet (Manchester: W. E. Clegg, 1894).

4 *"Proconsul africanus":* A. T. Hopwood, "Miocene Primates from Kenya," *Journal of the Linnean Society London* 38 (1933): 437–464.

5 "That Consul lived offstage": Anonymous, "Le chimpanzee," 416.

5 "He has suggested": D. E. Wildman, M. Uddin, G. Liu, L. I. Grossman, and M. Goodman, "The Role of Natural Selection in Shaping 99.4% Identity between Humans and Chimpanzees at Nonsynonymous DNA Sites: Implications for Enlarging the Genus *Homo*," *Proceedings of the National Academy of Science* 100 (2003): 7181–7188.

1 Luck and Unluck

11 "Dayrell Botry Pigott": Contrary to "Prehistoric Mammals; Fossil-Hunter's Labours," *Daily Telegraph,* April 5, 1912, the unlucky Mr.

Pigott was named Dayrell Botry Pigott, not Digby, a misunderstanding that has clung to his story ever since. See "Mr. D. B. Pigott," *The Times,* March 9, 1911, 11, for an obituary.

12 "He was sent out": P. J. Andrews, "A Short History of Miocene Field Paleontology in Western Kenya," *Journal of Human Evolution* 10 (1981): 3–9.

12 "Hobley was something": A.-T. Matson and T. F. Ofcansky, "A Bio-Bibliography of C. W. Hobley," *History in Africa* 8 (1981): 253–260; C. W. Hobley, *Kenya: From Chartered Company to Crown Colony.* (London: Frank Cass, 1970).

12 "D. B. Pigott's family history": J. A. Venn, *Alumni Cantabrigienses: A Biographical List of All Known Students, Graduates, and Holders of Office at the University of Cambridge from the Earliest Times to 1900* (Cambridge: Cambridge University Press, 1953), 124.

13 "He'd been in that post": "Pigott," *The Times,* 11.

13 "Pigott duly collected": F. Oswald, *Alone in Sleeping-Sickness Country* (London: Kegan, Paul, Trench, 1915).

14 "By November": F. Oswald, Correspondence and Notes concerning the Oswald Expedition to Victoria Nyanza, Archives NHM, Department of Palaeontology (1911); "Prehistoric," *Daily Telegraph,* 1912.

14 "On February 28, 1911": "Pigott," *The Times,* 11.

15 "Hobley records that the duke": Hobley, *Kenya,* 207; Oswald, *Alone,* 5.

15 "by drowning"; "His body": "Pigott," *The Times,* 11.

15 "*Crocodylus pigotti*": E. Tchernov and J. Van Couvering, "New Crocodiles from the Early Miocene of Kenya," *Paleontology* 21 (1978): 857–867.

16 "More Miocene fossils": Andrews, "Short History," 3–9.

16 "Edward James Wayland": K. A. Davies, "E. J. Wayland, C.B.E.—A Tribute," *Uganda Journal* 31 (1967): 1–3; Lord Teining, "E. J. Wayland and the Founding of the Uganda Society," *Uganda Journal* 31 (1967): 3–4.

17 "It is with much regret": E. J. Wayland, *Annual Report of the Geological Survey of Uganda for the Year Ending 31st March, 1920* (Entebbe: Entebbe Government Printer, 1920), 6–7.

17 "remarkable lack of interest": Ibid., 8.

18 "Wayland looked him straight in the eye": J. Sykes, "Wayland and the Uganda Journal," *Uganda Journal* 31 (1967): 4–5.

18 “Hopwood was a very tall”: H. B. S. Cooke, letter to Pat Shipman, Oct. 16, 2000.

18 “*Proconsul,* to almost anyone”: Hopwood, “Miocene Primates,” 438.

18 “Louis was legendary”: Information from the authors’ personal knowledge; L. S. B. Leakey, *White African* (London: Holder and Stoughton, 1937; reprinted Cambridge: Schenkman, 1966); V. Morell, *Ancestral Passions: The Leakey Family and the Quest for Humankind’s Beginnings* (New York: Touchstone Press, 1996).

20 “In the five weeks”: Hopwood, “Miocene Primates,” 437.

21 “On September 22”: Leakey, *White,* 283.

21 “Louis had already publicly proclaimed”: L. S. B. Leakey, “Earliest Man in East Africa,” *The East African Standard,* Oct. 10, 1931; L. S. B. Leakey, “Earliest Man, Discoveries in East Africa,” *The Times,* March 9, 1932, 11.

21 “Reck had been able”: Morell, *Ancestral,* 59.

21 “The key question was”: Morell, *Ancestral,* 54–56.

21 “Sir Arthur Keith”: A. Keith, *New Discoveries Relating to the Antiquity of Man* (London: Norgate, 1931), 155, 172.

22 “Louis was deeply annoyed”: Morell, *Ancestral,* 64, citing J. Solomon to V. Morell, March 30, 1985; L. S. B. Leakey to A. T. Hopwood, March 18, 1932, Archives NHM.

22 “I WANT IF POSSIBLE”: Morell, *Ancestral,* 64, citing L. S. B. Leakey to A. Keith, Feb. 26, 1932.

23 “As Louis’s colleague”: Morell, *Ancestral,* 69 citing J. Solomon to V. Morell, Oct. 30, 1985.

23 “In 1932, Louis announced”: L. S. B. Leakey, “The Oldoway Human Skeleton,” *Nature* 129 (1932): 721–722; L. S. B. Leakey, “The Oldoway Human Skeleton,” *Nature* 130 (1932), 578; L. S. B. Leakey, *Stone Age Races of Kenya* (London: Oxford University Press, 1935), 11–23.

23 “not able to point to any”; “congratulated on the . . . exceptional”: Anonymous, “Early Human Remains in East Africa: Report of a Conference Convened by the Royal Anthropological Institute at Cambridge,” *Man* 65 (1933): 210.

24 “found Louis personally irritating”: Morell, *Ancestral,* 86–87.

24 “ten yards or so”: L. S. B. Leakey, field diary, Jan. 16, 1935, Archives KNM.

24 “When Boswell’s assessment”: P. G. H. Boswell, “Human Remains from Kanam and Kanjera, Kenya Colony,” *Nature* 135 (1935): 371.

24 "He was passionately in love": Morell, *Ancestral,* 87–91.

25 "Louis's failure": Ibid.

25 "So far as I can gather": Morell, *Ancestral,* 91, citing A. C. Haddon to L. S. B. Leakey, March n.d., 1935.

25 "Boswell's findings may ruin": L. S. B. Leakey, field diary, March 16, 1935, Archives KNM.

25 "The paper that Boswell": Boswell, "Human Remains," 371; see also L. S. B. Leakey, "Fossil Human Remains from Kanam and Kanjera, Kenya Colony," *Nature* 138 (1936): 643.

26 "In all, they recovered": L. S. B. Leakey, "East African Archaeological Expedition, Fourth Season 1934–35, Fourth Monthly Report, Jan. 24–Feb. 23," Archives NHM.

26 "In the normal course": Leakey, *White,* 311.

26 "I am very sorry": A. T. Hopwood to L. S. B. Leakey, March 18, 1935, Archives KNM.

27 "Sometimes Louis wrote up": L. S. B. Leakey, "A Miocene Anthropoid Mandible from Rusinga, Kenya," *Nature* 152 (1943): 319–20; D. MacInnes, "Notes on the East African Miocene Primates," *Journal of the East African Natural History Society* 17 (1943): 141–181.

2 Love and the Tree

28 "Mary Nicol arrived": M. D. Leakey, *Disclosing the Past* (Garden City: Doubleday, 1984), 49.

28 "What the hell": V. Morell, *Ancestral Passions: The Leakey Family and the Quest for Humankind's Beginnings* (New York: Touchstone Press, 1996), 97.

28 "White recalled that": Morell, *Ancestral,* 99.

29 "as her quiet competence": Ibid.

29 "His fellowship was not": Ibid., 105.

30 "As a finale": Ibid., 141.

31 "From 1925, when Dart": R. A. Dart, "*Australopithecus africanus:* The Man-Ape of South Africa," *Nature* 115 (1925): 195–199.

32 "Broom, a stubborn old Scots physician": R. A. Dart with D. Craig, *Adventures with the Missing Link* (New York: Harper & Bros, 1959), 47; R. Broom, *Finding the Missing Link* (London: C. A. Watts, 1950).

32 "australopithecines were largely dismissed": see, for example, A. Keith,

G. E. Smith, A. S. Woodward, and W. J. H. Duckworth, "The Fossil Anthropoid from Taungs," *Nature* 116 (1925): 11; A. Keith, "The Taungs Skull," *Nature* 116 (1925): 462.

32 "Le Gros announced his radical conclusion ": W. E. L. G. Clark, "Observations on the Anatomy of the Fossil Australopithecinae," *Journal of Anatomy* 81 (1947): 300–333; "'Missing Link' Found in Africa," *New York Times,* Jan. 23, 1947, 25; A. Keith, "Australopithecinae or Dartians," *Nature* 159 (1947): 377.

32 "As early as 1871": C. Darwin, *Descent of Man* (New York: A. L. Burt, 1871), 176–177.

33 "She had some difficulty": L. S. B. Leakey, *By the Evidence: Memoirs, 1932–1951* (New York: Harcourt, Brace, 1974), 207.

33 "was put gently but firmly": Leakey, *Disclosing,* 94.

33 "Many of the islanders": Ibid.

34 "Le Gros left the conference": Morell, *Ancestral,* 143.

34 "Within weeks, Le Gros had": Ibid., 145.

34 "Phillips had raised": W. Phillips to L. S. B. Leakey, Sept. 2, 1947, Archives KNM.

35 "Louis was greatly relieved": L. S. B. Leakey to W. E. L. G. Clark, Dec. 29, 1947, Archives KNM.

35 "The Maji Moto": Morell, *Ancestral,* 146.

36 "The British-Kenyan Miocene Expedition": Leakey, *Disclosing,* 97ff.

37 "I would never claim": Ibid., 96.

37 "never cared . . . for crocodiles": Ibid., 98.

39 "She shouted to attract": Morell, *Ancestral,* 150.

39 "This was a wildly exciting find ": Leakey, *Disclosing,* 98–99.

40 "She and Louis celebrated": Ibid., 99.

40 "WE GOT THE BEST PRIMATE": L. S. B. Leakey to W. E. L. G. Clark, Oct. 10, 1948, Archives KNM.

40 "She was greeted by": M. D. Leakey to L. S. B. Leakey, Oct. 31, 1948, Archives KNM.

41 "resemblances to the human condition"; "To me, this particular": L. S. B. Leakey to W. E. L. G. Clark, Nov. 8, 1948, Archives KNM, Rus/LSBL 256.

42 "there necessarily remains an element of doubt": W. E. L. G. Clark and L. S. B. Leakey, *The Miocene Hominoidea of East Africa,* Fossil Mam-

mals of Africa, vol.1, 1–117 (London: British Museum of Natural History, 1951), 3.

42 "While there seems no reason": Clark and Leakey, *Miocene,* 5–6.

42 "We have not attempted": Ibid., 10.

43 "not one but six species": W. E. L. G. Clark and L. S. B. Leakey, "Diagnoses of East African Miocene Hominoidea," *Quarterly Journal of the Geological Society of London* 105 (1950): 260–262.

43 "MacInnes, who collected the specimen": D. G. MacInnes, "Notes on East African Miocene Primates," *Journal of the East African Natural History Society* 17 (1943): 141–181.

43 "An analysis of the matrix": L. S. B. Leakey, "Notes on the Mammalian Faunas from the Miocene and Pleistocene of East Africa," in W. W. Bishop and J. D. Clark, ed., *Background to Evolution in Africa* (Chicago: University of Chicago Press, 1967), 7–29; P. Andrews and T. Molleson, "The Provenance of *Sivapithecus africanus,*" *Bulletin of the British Museum (Natural History) Geological Series,* 32 (1979):19–32.

44 "It seems possible": Clark and Leakey, *Miocene,* 6–7.

44 "the anecdote was a favorite": J. Van Couvering to Alan Walker, 2004.

44 "We now know that": M. Pickford, "Sedimentation and Fossil Preservation in the Nyanza Rift Valley System, Kenya," in L. S. Frostick, R. W. Renaut, I. Reid, and J. J. Tiercelin, ed., *Sedimentation in the African Rifts,* Geological Society Special Publications n. 25 (London, 1986), 345–362.

45 "intermediate in size"; "species of *Proconsul*": Clark and Leakey, *Miocene,* 11–12.

46 "three species of chimpanzees": P. A. Morin, J. Moore, R. Chakraborty, L. Jin, J. Goodall, and D. S. Woodruff, "Kin Selection, Social Structure, Gene Flow, and the Evolution of Chimpanzees," *Science* 265 (1994): 1193–1201; Conservation International, "Once and Future Primate Order," press release, April 5, 2000.

51 "If *Proconsul* occasionally raised itself": Clark and Leakey, *Miocene,* 92.

3 An Arm and a Leg

53 "Knowing that more fossils": P. Andrews, "A Short History of Miocene Field Paleontology in Western Kenya," *Journal of Human Evolution* 10 (1981): 3–9.

53 "a fossilized fig with a bite mark": P. Davis to A. Walker, Aug. 29, 1991.

55 "Ol Doinyo Lengai": R. L. Hay and R. J. Reeder, "Calcretes of Olduvai Gorge and the Ndolanya Beds of Northern Tanzania," *Sedimentology* 25 (1978): 649–673; J. K. Bourne, "Ol Doinyo Lengai," *National Geographic Magazine* 203 (2003): 34–49.

55 "the fossil soils at Rusinga": G. J. Retallack, E. A. Bestland, and D. P. Dugas, "Miocene Paleosols and Habitats of *Proconsul* on Rusinga Island, Kenya," *Journal of Human Evolution* 29, no. 1 (1995): 53–91.

55 "the mock aridity in the soils": J. A. Harris and J. A. Van Couvering, "Mock Aridity and the Paleoecology of Volcanically Influenced Ecosystems," *Geology* 23, no. 4 (1995): 593–596.

55 "The prodigious amounts": M. Pickford, "Sedimentation and Fossil Preservation in the Nyanza Rift Valley System, Kenya," in L. S. Frostick, R. W. Renaut, I. Reid, and J. J. Tiercelin, ed., *Sedimentation in the African Rifts,* Geological Society Special Publications no. 25 (London, 1986), 345–362.

56 "Another stunning set of fossil finds": T. Whitworth, "A Contribution to the Geology of Rusinga Island, Kenya," *Quarterly Journal of the Geological Society* 105, no. 1 (1953):75–96.

56 "A later tally": A. Walker and M. Pickford, "New Postcranial Fossils of *Proconsul africanus* and *P. nyanzae,*" in R. E. Ciochon and R. S. Corrucini, ed., *New Interpretations of Ape and Human Ancestry* (New York: Plenum Press, 1983), 325–352.

56 "The profusion of articulated skeletons": Whitworth, "Contribution," 91.

57 "Within a few months": W. E. L. G. Clark, *Chant of Pleasant Exploration* (London: Livingstone, 1968), 134–135.

57 "overbearing"; "autocratic manner"; "clearly-expressed"; "merciless attacks": J. Peyton, *Solly Zuckerman: A Scientist Out of the Ordinary* (London: John Murray, 2001), pp. 32–33.

58 "Zuckerman went to Birmingham": Ibid..

58 "a series of harsh papers": S. Zuckerman, "South African Anthropoids," *Nature* 166 (1950): 188; E. Ashton and S. Zuckerman, "Some Quantitative Dental Characters of Fossil Anthropoids," *Philosophical Transactions of the Royal Society London* B, 234 (1950): 485–520; E. Ashton and S. Zuckerman, "Statistical Methods in Anthropology," *Nature* 168 (1951): 1116.

58 "Hardly one of the [australopithecine] teeth": Ashton and Zuckerman, "Quantitative," 520.

59 "when two statisticians pointed out": F. Yates and M. T. R. Healey, "Statistical Methods in Anthropology," *Nature* 168 (1951): 1116; J. Bronowski and and W. M. Long, "Statistical Methods in Anthropology, *Nature* 168 (1951): 794.

60 "he felt a new age was dawning": J. R. Napier, personal communication to Alan Walker.

60 "John Napier": Information on John Napier from the personal knowledge of Alan Walker; L. Aiello, "Preface," *Journal of Human Evolution* 22 (1992): 239–243; Anonymous, "Obituary: John Russell Napier," *Journal of Human Evolution* 16 (1987): 533–535; A. Walker, "Louis Leakey, John Napier, and the History of *Proconsul*," *Journal of Human Evolution* 22 (1992): 245–254.

61 "the use of the forelimbs": W. E. L. G. Clark, "Preface," in J. R. Napier and P. Davis, *The Forelimb Skeleton and Associated Remains of* Proconsul africanus," Fossil Mammals of Africa, 16 (London: British Museum [Natural History], 1959), v.

61 "*vertical clinging and leaping*": J. R. Napier and A. Walker, "Vertical Clinging and Leaping—a Newly Recognized Category of Locomotor Behaviour of Primates," *Folia Primatologica* 6 (1967): 204–219.

61 "It is clear": Napier and Davis, *Forelimb,* 1.

64 "shows many primitive": Ibid., 56.

64 "the paleobotanist Katherine Chesters": K. I. M. Chesters, "The Miocene Flora of Rusinga Island, Lake Victoria, Kenya," *Paleontographica* section B 101 (1957): 30–71.

64 "The importance of the conclusion": Napier and Davis, *Forelimb,* 62–63.

65 "I have no doubt": W. E. L. G. Clark to D. Allbrook, Nov. 20, 1962.

67 "slightly crazy character named Ike Russell": See T. Bowen, ed., *Backcountry Pilot: Flying with Ike Russell* (Tucson: University of Arizona Press, 2002).

69 "John collected the first samples": J. A. Van Couvering and J. A. Miller, "Miocene Stratigraphy and Age Determinations, Rusinga Island, Kenya," *Nature* 221 (1969): 628–632.

70 "This age was confirmed": R. E. Drake, J. A. Van Couvering, M.

Pickford, G. H. Curtis and J. A. Harris, "New Chronology of the Early Miocene Mammalian Faunas of Kisingiri, Western Kenya," *Journal of the Geological Society of London* 145 (1988): 479–491.

70 "The authors of the original study": W. W. Bishop, J. A. Miller, and F. J. Fitch, "New Potassium-Argon Determinations Relevant to Miocene Fossil Mammal Sequence in East Africa," *American Journal of Science* 267 (1969): 669–699.

4 The Lost and the Found

73 "With all these experts": M. Pickford to A. Walker, July 14, 1981.

74 "Martin and I published a preliminary paper": A. Walker and M. Pickford, "New Postcranial Fossils of *Proconsul africanus* and *P. nyanzae*," in R. E. Ciochon and R. S. Corrucini, ed., *New Interpretations of Ape and Human Ancestry* (New York: Plenum Press, 1983), 325–352.

74 "When we found the complete right metacarpal": K. C. Beard, M. F. Teaford, and A. Walker, "New Wrist Bones from *Proconsul africanus* and *P. nyanzae* from Rusinga Island, Kenya," *Folia Primatologica* 47 (1986): 97–118.

76 "The exact nature": P. Andrews, "A Short History of Miocene Field Paleontology in Western Kenya," *Journal of Human Evolution* 10 (1981): 7.

76 *August 27th, 1951:* Thomas Whitworth, field diary, 1951, Archives KNM.

77 *December 16, 1951*: L. S. B. Leakey, field diary, 1951, Archives KNM.

78 "He wondered if the 'turtle scutes'": A. Walker, "The Puzzle of *Proconsul*," *The Sciences* 23 (1983): 23; A. Walker and M. F. Teaford, "The Hunt for *Proconsul*," *Scientific American* 260, no. 1 (1989): 76–78; A. Walker, D. Falk, R. Smith, and M. Pickford, "The Skull of *Proconsul africanus:* Reconstruction and Cranial Capacity," *Nature* 305 (1983): 525.

78 "Though I'd at first assumed": Walker and Teaford, "Hunt," 76.

78 "When Richard Leakey, Louis's son, became": Richard became administrative director in 1968.

79 "may be housed": acting chief secretary of the Kenya government to Louis Leakey, Nov. 21, 1981, Archives KNM.

79 "Documents in the Natural History Museum's archives": File De-Registration, 1981 DIR 39, Archives NHM, including: H. W. Ball to R. E. Leakey, July 23, 1981; S. Runyard to H. W. Ball, June 30, 1981; Formal Statement of De-Registration 22.7.81.

79 "The newspapers and magazines": headlines quoted in File De-Registration, 1981 DIR 39, Archives NHM.

80 "she had said in print": D. Falk, "A Reconsideration of the Endocast of *Proconsul africanus*," in Ciochon and Corrucini, *New,* 239–248.

81 "Together, we attacked": Walker et al., "Skull," 525–527.

83 "*Proconsul* had an EQ": Ibid., 527. The equation we used was EQ = {(brain weight in grams) divided by [(0.0991) multiplied by (body weight in grams)$^{0.76237}$]}, from R. L. Holloway and D. G. Post, "The Relativity of Relative Brain Measures," in E. Armstrong and D. Falk, ed., *Primate Brain Evolution* (New York: Plenum, 1981), 60. The result of this equation was divided by 2.87 to convert the answer into a percentage of the human EQ.

84 "gorillas have the highest": Holloway and Post, "Relativity," 60.

84 "The question of diet": R. D. Martin, *Human Brain Evolution in an Ecological Context,* 52nd James Arthur Lecture (New York: American Museum of Natural History, 1983); T. Clutton-Brock and P. Harvey, "Primates, Brains, and Ecology," *Journal of Zoology* 190 (1980): 309–323.

85 "In 1997, the discovery of ": B. R. Benefit and M. L. McCrossin, "Earliest Known Old World Monkey Skull," *Nature* 388 (1997): 368–371.

86 "By using the postcrania": T. Harrison, "New Postcranial Remains of *Victoriapithecus* from the Middle Miocene of Kenya," *Journal of Human Evolution* 18 (1989): 3–54.

5 Back to the Miocene

87 "When Richard was a child": The following anecdote is from V. Morell, *Ancestral Passions: The Leakey Family and the Quest for Humankind's Beginnings* (New York: Touchstone Press, 1996), 157.

90 "We left Nairobi": The account that follows and direct quotations not otherwise attributed are from A. Walker, field diary, 1984.

111 "Our scenario was this": A. Walker and M. F. Teaford, "The Hunt for *Proconsul,*" *Scientific American* 260, no. 1 (1989): 76–78.

6 An Embarrassment of Riches

113 "We were back at Rusinga": The account that follows and quotations not otherwise attributed are taken from A. Walker, field diaries, 1984, 1985, 1986, 1987.

115 “fills the air”: F. Oswald, *Alone in Sleeping Sickness Country* (London: Kegan, Paul, Trench, 1915), 81.

119 “Later I saw an engraving”: S. Baker, *The Nile Tributaries of Abyssinia* (London: Macmillan, 1886), facing p. 172.

132 “the first pelvis of any decent Miocene ape”: The first brief note about this specimen was C. V. Ward, A. Walker, M. Teaford, and I. Odhiambo, “*Proconsul nyanzae* Innominate from the Early Miocene of Mfangano Island, Kenya,” *American Journal of Physical Anthropology* 78 (1989): 319–320.

132 “most complete *Proconsul nyanzae* known”: C. V. Ward, M. Teaford, A. Walker, and I. Odhiambo, “Partial Skeleton of *Proconsul nyanzae* from Mfangano Island, Kenya,” *American Journal of Physical Anthropology* 90 (1993): 77–111.

133 “By the end of our last field season”: A. Walker and M. Teaford, “The Kaswanga Primate Site: An Early Miocene Hominoid Site on Rusinga Island, Kenya,” *Journal of Human Evolution* 17 (1988): 539–544.

7 How Did It Move?

138 “*Proconsul* Did Not Have a Tail”: C. V. Ward, A. Walker, and M. Teaford, “*Proconsul* Did Not Have a Tail,” *Journal of Human Evolution* 21 (1991): 215–220.

141 “Tail loss is a diagnostic feature”: Ibid., 220.

141 “Later, Terry Harrison”: T. Harrison, “Evidence for a Tail in *Proconsul heseloni*,” *American Journal of Physical Anthropology* Supplement 26 (1998): 93–94.

142 “The Samburu fossils”: H. Ishida, M. Pickford, H. Nakaya, and Y. Nakano, “Fossil Anthropoids from Nachola and Samburu Hills, Samburu District, Northern Kenya,” *African Study Monographs* Supplementary issue 2 (1984): 73–86; M. D. Rose, Y. Nakano, and H. Ishida, “*Kenyapithecus* Postcranial Specimens from Nachola, Kenya,” *African Study Monographs* Supplementary issue 24 (1996): 3–56; M. Nakatsukasa, A. Tamanaka, Y. Kunimatsu, D. Shimuzu, and H. Ishida, “A Newly Discovered *Kenyapithecus* Skeleton and Its Implications for the Evolution of Positional Behavior in Miocene East African Hominoids,” *Journal of Human Evolution* 34 (1998) 657–664; H. Ishida, Y. Kunimatsu, M. Nakatsukasa, and Y. Nakano, “New Hominoid Ge-

nus from the Middle Miocene of Nachola, Kenya," *Anthropological Science* 107, no. 2 (1999): 189–191; M. Nakatsukasa, Y. Kunimatsu, Y. Nakano, and H. Ishida, "A New Skeleton of the Large Hominoid from Nachola, Northern Kenya," *American Journal of Physical Anthropology* Supplement 30 (2000): 235.

142 "an unusually long forelimb": T. Takano, M. Nakatsukasa, Y. Kunimatsu, Y. Nakano, and H. Ishida, "Functional Morphology of the *Nacholapithecus* Forelimb Long Bones," *American Journal of Physical Anthropology* Supplement 36 (2003): 205–206.

142 "nineteen partial . . . vertebrae": T. Takano, M. Nakatsukasa, Y. Kunimatsu, Y. Nakano, and H. Ishida, "Morphology of the Axial Skeleton of *Nacholapithecus* from the Middle Miocene of Kenya," *American Journal of Physical Anthropology Supplement* 36 (2003): 206.

142 "*Nacholapithecus* lacked a tail": M. Nakatsukasa, H. Tsujikawa, D. Shimuzu, T. Takano, Y. Kunimatsu, Y. Nakano, and H. Ishida, "Definitive Evidence for Tail Loss in *Nacholapithecus,* an East African Miocene Hominoid," *Journal of Human Evolution* 45 (2003): 179–186.

143 "*Proconsul* cannot have borne a tail": M. Nakatsukasa, C. V. Ward, A. Walker, M. Teaford, and N. Ogihara, "Tail Loss in *Proconsul heseloni,*" *Journal of Human Evolution,* 46 (2004): 777–784.

143 "For her Ph. D. thesis": C. V. Ward, "The Functional Anatomy of the Lower Back and Pelvis of the Miocene Hominoid *Proconsul nyanzae* from Mfangano Island, Kenya" (Ph.D. diss., The Johns Hopkins University School of Medicine, 1991); C. V. Ward, "Partial Skeleton of *Proconsul nyanzae* from Mfangano Island, Kenya," *American Journal of Physical Anthropology* 90 (1993): 77–111; C. V. Ward, "Torso Morphology and Locomotion in *Proconsul nyanzae,*" *American Journal of Physical Anthropology* 92 (1993): 291–328.

149 "Old World monkeys . . . terrestrial": See M. L. McCrossin, B. R. Benefit, S. N. Gitau, A. K. Palmer, and K. T. Blue, "Fossil Evidence for the Origins of Terrestriality among Old World Higher Primates," in E. Strasser, J. G. Fleagle, A. Rosenberger, and H. McHenry, ed., *Primate Locomotion: Recent Advances* (New York: Plenum, 1998), 353–396.

149 "fossil kneecaps from Rusinga": C. V. Ward, C. B. Ruff, A. Walker, M. Teaford, M. D. Rose, and I. O. Nengo, "Functional Morphology of

Proconsul Patellas from Rusinga Island, Kenya," *Journal of Human Evolution* 29 (1995): 1–19.

151 "much of the skeleton of *Proconsul nyanzae*": Ward, "Torso."

152 "vertebrae attributed to . . . *Proconsul major*": A. Walker and M. D. Rose, "Fossil Hominoid Vertebrae from the Miocene of Uganda," *Nature* 217 (1968): 980–981.

152 "the Moroto vertebrae": R. E. Leakey and M. G. Leakey, "A New Miocene Hominoid from Kenya," *Nature* 324 (1986): 143–145; R. E. Leakey, M. G. Leakey, and A. Walker, "Morphology of *Afropithecus turkanensis* from Kenya," *American Journal of Physical Anthropology* 76 (1988): 289–307; Ward, "Torso."

8 How Many *Proconsuls*?

153 "*Proconsul africanus* was first named": A. T. Hopwood, "Miocene Primates from Kenya," *Journal of the Linnean Society London* 38 (1933): 437–464.

154 "In 1950 and 1951": W. E. L. G. Clark and L. S. B. Leakey, "Diagnoses of East African Miocene Hominoidea," *Quarterly Journal of the Geological Society of London* 105 (1950): 260–262; W. E. L. G. Clark and L. S. B. L. Leakey, *The Miocene Hominoidea of East Africa,* Fossil Mammals of Africa, vol. 1, 1-117 (London: British Museum [Natural History], 1951).

154 "In 1968, Mike Rose and I": A. Walker and M. D. Rose, "Fossil Hominoid Vertebrae from the Miocene of Uganda," *Nature* 217 (1968): 980–981.

154 "In 1965, Elwyn Simons and David Pilbeam": E. L. Simons and D. R. Pilbeam, "Preliminary Revision of the Dryopithecinae," *Folia Primatologica* 3 (1965): 1–152.

154 "David speculated that *P. major*": D. R. Pilbeam, "Tertiary Pongidae of East Africa: Evolutionary Relationships and Taxonomy," *Bulletin of the Peabody Museum of Natural History* 31 (1969): 1–185.

155 "Leonard Greenfield noted": L. Greenfield, "Sexual Dimorphism in *Dryopithecus africanus*," *Primates* 13 (1972): 395–420.

155 "Wendy Bosler, went a step further": W. Bosler, "Species Groupings of Early Miocene Dryopithecine Teeth from East Africa," *Journal of Human Evolution* 10 (1981): 151–158.

155 "Might the two 'species'": J. Kelley, "Species Recognition and Sexual Dimorphism in *Proconsul* and *Rangwapithecus*," *Journal of Human Evolution* 1015 (1986): 461–495; M. Pickford, "Sexual Dimorphism in *Proconsul*," in M. Pickford and B. Chiarelli, ed., *Sexual Dimorphism in Living and Fossil Primates* (Florence: Il Sedicesimo, 1986), 133–170.

155 "the principle of uniformitarianism": For cogent discussions of uniformitarianism and its application, see J. Kelley, "Taxonomic Implications of Sexual Dimorphism in *Lufengpithecus*," in W. H. Kimbel and L. B. Martin, ed., *Species, Species Concepts, and Primate Evolution* (New York: Plenum, 1993), 429–458, and M. Teaford, A, Walker, and G. S. Mugaisi, "Species Discrimination in *Proconsul* from Rusinga and Mfangano Islands, Kenya," in Kimbel and Martin, ed., *Species*, 373–392.

156 "Orangutans are the most dimorphic": C. Oxnard, S. S. Lieberman, and B. R. Gelvin, "Sexual Dimorphism in Dental Dimension of Higher Primates," *American Journal of Physical Anthropology* 8 (1985): 127–152.

157 "Writing with David Pilbeam": J. Kelley and D. R. Pilbeam, "The Dryopithecines: Taxonomy, Anatomy, and Phylogeny of the Miocene Large Hominoids," in D. R. Swinder and J. Erwin, ed., *Comparative Primate Biology,* vol. 1*: Systematics, Evolution, and Anatomy* (New York: Alan R. Liss, 1986), 361–411; Kelley, "Taxonomic Implications."

157 "To point up this problem": M. Teaford, K. C. Beard, R. E. Leakey, and A. Walker, "New Hominoid Facial Skeleton from the Early Miocene of Rusinga Island, Kenya, and Its Bearing on the Relationship between *Proconsul nyanzae* and *Proconsul africanus*," *Journal of Human Evolution* 17 (1988): 461–477.

159 "In skeletons of hominoid species": H. McHenry, "Size Variation in the Postcranium of *Australopithecus afarensis* and Extant Species of Hominoidea," *Human Evolution* 1 (1986): 149–156.

159 "The wrist bones": K. C. Beard, M. Teaford, and A. Walker, "New Wrist Bones from *Proconsul africanus* and *P. nyanzae* from Rusinga Island, Kenya," *Folia Primatologica* 47 (1986): 97–118.

159 "the large bones of the ankle": Teaford, Walker, and Mugaisi, "Species Discrimination."

161 "In the initial phase of this study": C. B. Ruff, A. Walker, and M. Teaford, "Body Mass, Sexual Dimorphism, and Femoral Proportions of *Proconsul* from Rusinga and Mfangano Islands, Kenya," *Journal of Human Evolution* 18 (1989): 515–536.

162 "'Such extreme variation"; "It therefore appears most likely": Ibid., 529–530.

163 "She wanted to determine": K. Rafferty, A. Walker, C. B. Ruff, M. D. Rose, and P. Andrews, "Postcranial Estimates of Body Weight in *Proconsul,* with a Note on a Distal Tibia of *P. major* from Napak, Uganda," *American Journal of Physical Anthropology* 97 (1995): 391–402.

164 "Even Jay Kelley came around": J. Kelley, "Life-History Evolution in Miocene and Extant Apes," in N. Minugh-Purvis and K. J. McNamara, ed., *Human Evolution through Developmental Change* (Baltimore: The Johns Hopkins University Press, 2002), 236.

164 "In 1993, four of us undertook . . . a virtual collaboration": A. Walker, M. Teaford, L. Martin, and P. Andrews, "A New Species of *Proconsul* from the Early Miocene of Rusinga/Mfangano Islands, Kenya," *Journal of Human Evolution* 25 (1993): 43–56.

165 "In his correspondence with Le Gros": L. S. B. Leakey to W. E. L. G. Clark, Nov. 8, 1948, RUS/LSBL 256, Archives KNM; L. S. B. Leakey to W. E. L. G. Clark, Oct. 10, 1948, NAK/MUS/1/693 Rusinga 1947–48, Archives KNM.

9 How Many Apes?

179 "this new Buluk creature": R. E. Leakey and A. Walker, "New Higher Primates from the Early Miocene of Buluk, Kenya," *Nature* 318 (1985): 173–175.

179 *"Afropithecus turkanensis":* R. E. Leakey, M. G. Leakey, and A. Walker, "Morphology of *Afropithecus turkanensis* from Kenya," *American Journal of Physical Anthropology* 76 (1988): 289–307; M. G. Leakey and Alan Walker, "*Afropithecus:* Function and Phylogeny," in D. R. Begun, C. V. Ward, and M. D. Rose, ed., *Function, Phylogeny, and Fossils: Miocene Hominoid Evolution and Adaptations* (New York: Plenum. 1997), 225–239.

181 *"Aegyptopithecus zeuxis":* M. G. Leakey, R. E. Leakey, J. T. Richtsmeier,

E. L. Simons, and A. Walker, "Similarities in *Aegyptopithecus* and *Afropithecus* Facial Morphology," *Folia Primatologica* 56, no. 2 (1991): 65–85.

181 "*Aegyptopithecus* is from the early Oligocene": J. G. Fleagle, T. M. Brown, J. D. Obradovich, and E. L. Simons, "How Old Are the Fayum Primates?" In J. B. Else and P. Lee, ed., *Primate Evolution* (Cambridge: Cambridge University Press, 1986), 3–17.

181 "it was probably a seed predator": See, e.g., M. G. M. van Roosmalen, R. A. Mittermaier, and J. G. Fleagle, "Diet of the Northern Bearded Saki *(Chiropotes satanas chiropotes):* A Neotropical Seed Predator," *American Journal of Primatology* 14 (1988): 11–35.

182 "tapirs are well known to be seed predators": D. H. Jansen, "Digestive Seed Predation by a Costa Rican Baird's Tapir," *Biotropica* Supplement 13 (1981): 59–63; M. Rodrigues, F. Olmos, and M. Galetti, "Seed Dispersal by Tapir in Southeastern Brazil," *Mammalia* 57, no. 3 (1993): 460–461.

182 "*Turkanapithecus kalakolensis*": R. E. Leakey and M. G. Leakey, "A New Miocene Hominoid from Kenya," *Nature* 324 (1986): 143–146; R. E. Leakey, M. G. Leakey, and A. Walker, "Morphology of *Turkanapithecus kalakolensis* from Kenya," *American Journal of Physical Anthropology* 76 (1988): 277–288.

182 "*Simiolus enjiessi*": M. D. Rose, M. G. Leakey, R. E. Leakey, and A. Walker, "Postcranial Remains of *Simiolus enjiessi* and Other Primitive Catarrhines from the Early Miocene of Lake Turkana, Kenya," *Journal of Human Evolution* 22 (1992): 171–237.

188 "David Begun's work on the phalanges": D. R. Begun, M. Teaford, and A. Walker, "Comparative Functional Anatomy of the *Proconsul* Phalanges from the Kaswanga Primate Site, Rusing, Kenya," *Journal of Human Evolution* 26 (1994): 89–145.

189 "stem catarrhines": T. Harrison, "The Phylogenetic Relationships of the Early Catarrhine Primates: A Review of the Current Evidence," *Journal of Human Evolution* 16 (1987): 41–80.

190 "*Kenyapithecus wickeri*": L. S. B. Leakey, "A New Lower Pliocene Fossil Primate from Kenya," *Annals and Magazine of Natural History* 4 (1962): 689–696; L. S. B. Leakey, "The Lower Dentition of *Kenyapithecus wickeri*," *Nature* 217 (1968): 827–830.

191 *"Kenyapithecus africanus":* M. Nakatsukasa, A. Tamanaka, Y. Kunimatsu, D. Shimuzu, and H. Ishida, "A Newly Discovered *Kenyapithecus* Skeleton and Its Implications for the Evolution of Positional Behavior in Miocene East African Hominoids," *Journal of Human Evolution* 34 (1998) 657–664; B. R. Benefit and M. L. McCrossin, "New *Kenyapithecus* Postcrania and Other Primate Fossils from Maboko Island, Kenya," *American Journal of Physical Anthropology* 16 (1993): 55–56; M. L. McCrossin and B. R. Benefit, "On the Relationships and Adaptations of *Kenyapithecus,* a Large-Bodied Hominoid from the Middle Miocene of Eastern Africa," in D. R. Begun, C. V. Ward, and M. D. Rose, ed., *Function, Phylogeny, and Fossils: Miocene Hominoid Evolution and Adaptations* (New York: Plenum, 1997), 241–67; C. S. Feibel and F. H. Brown, "Age of the Primate-Bearing Deposits on Maboko Island, Kenya," *Journal of Human Evolution* 21 (1991): 221–225.

191 "McCrossin's analysis": M. L. McCrossin, B. R. Benefit, S. N. Gitau, A. K. Palmer, and K. T. Blue, "Fossil Evidence for the Origins of Terrestriality among Old World Higher Primates," in E. Strasser, J. G. Fleagle, A. Rosenberger, and H. McHenry, ed., *Primate Locomotion: Recent Advances* (New York: Plenum, 1998), 353–396.

191 *"Nacholapithecus kerioi":* H. Ishida, Y. Kunimatsu, M. Nakatsukasa, and Y. Nakano, "New Hominoid Genus from the Middle Miocene of Nachola, Kenya," *Anthropological Science* 107, no. 2 (1999): 189–191; M. Nakatsukasa, Y. Kunimatsu, Y. Nakano, and H. Ishida, "A New Skeleton of the Large Hominoid from Nachola, Northern Kenya," *American Journal of Physical Anthropology* Supplement 30 (2000): 235.

192 *"Equatorius africanus":* S. Ward, B. Brown, A. Hill, J. Kelley, and W. Downs, "*Equatorius:* A New Hominoid Genus from the Middle Miocene of Kenya," *Science* 285 (1999): 1382–1386.

192 "a subfamily, the Afropithecinae": P. Andrews, "Evolution and Environment in the Hominoidea," *Nature* 360 (192): 641–647; S. Ward and D. Duren, "Middle and Late Miocene African Hominoids," in W. C. Hartwig, ed., *The Primate Fossil Record* (Cambridge: Cambridge University Press, 2002), 385–397.

192 "slowly through the trees on four feet": Leakey and Walker, *"Afropithecus."*

193 “four-footed ground dweller”: Ward et al., “*Equatorius.*”

193 “clambering and arm-swinging”: M. D. Rose, Y. Nakano, and H. Ishida, “*Kenyapithecus* Postcranial Specimens from Nachola, Kenya,” *African Study Monographs* Supplementary issue 24 (1996): 3–56.

193 “a strong link between *Kenyapithecus africanus* and *K. wickeri*”: B. R. Benefit and M. L. McCrossin, “Middle Miocene Hominoid Origins,” *Science* 287 (2000): 2375.

193 “*Equatorius* is the same species as *Griphopithecus*”: D. Begun, “Middle Miocene Hominoid Origins,” *Science* 287 (2000): 2375.

194 “What initially seemed”: Interview with Pat Shipman quoted in P. Shipman, “The Muddle in the Miocene,” *Anthroquest,* 46 (1992): 26.

194 “*Victoriapithecus* is also known”: Information on *Victoriapithecus* comes from B. R. Benefit, “The Permanent Dentition and Phylogenetic Position of *Victoriapithecus* from Maboko Island, Kenya,” *Journal of Human Evolution* 25 (1993): 83–172; B. R. Benefit “Phylogenetic, Paleodemographic, and Taphonomic Implications of *Victoriapithecus* Deciduous Teeth from Maboko, Kanya,” *American Journal of Physical Anthropology* 95 (1994): 277–331; B. R. Benefit and M. L. McCrossin, “Facial Anatomy of *Victoriapithecus* and Its Relevance to the Ancestral Cranial Morphology of Old World Monkeys and Apes,” *American Journal of Physical Anthropology* 92, no. 3 (1993): 329—370; B. R. Benefit and M. L. McCrossin, “The Victoriapithecidae, Cercopithecoidea,” in W. C. Hartwig, ed., *The Primate Fossil Record* (Cambridge: Cambridge University Press, 2002), 241–253; B. R. Benefit and M. L. McCrossin, “Earliest Known Old World Monkey Skull,” *Nature* 388 (1997): 368–371; T. Harrison, “New Postcranial Remains of *Victoriapithecus* from the Middle Miocene of Kenya,” *Journal of Human Evolution* 18 (1989): 3–54.

195 “*Prohylobates* is the only other monkey”: M. G. Leakey, “Early Miocene Cercopithecids from Buluk, Northern Kenya,” *Folia Primatologica* 44 (1985): 1–14; E. L. Simons, “Miocene Monkey *(Prohylobates)* from Northern Egypt, *Nature* 223 (1969): 687–689; E. Delson, “*Prohylobates* (Primates) from the Early Miocene of Libya: A New Species and Its Implication for Cercopithecid Origins,” *Geobios* 12 (1979): 725–733.

196 "the earliest monkeys specialized in habitats": Benefit and McCrossin, "Victoriapithecidae."

196 "environment has changed dramatically in Africa": T. E. Cerling, J. R. Ehleringer, and J. M. Harris, "Carbon Dioxide Starvation, the Development of C_4 Ecosystems, and Mammalian Evolution," *Philosophical Transactions of the Royal Society* B 353 (1998): 159–171; J. D. Kingston, B. D. Marin, and A. Hill, "Isotopic Evidence for Neogene Hominid Paleoenvironments in the Kenya Rift Valley," *Science* 264 (1994): 955–959.

196 "apes were common and monkeys were rare": N. Jablonski, "Fossil Old World Monkeys: The Late Neogene Radiation," in W. C. Hartwig, ed., *The Primate Fossil Record* (Cambridge: Cambridge University Press, 2002), 255–299; Benefit and McCrossin, "Victoriapithecidae"; T. Harrison, "Late Oligocene to Middle Miocene Catarrhines from Afro-Arabia," in Hartwig, ed., *The Primate Fossil Record,* 311–338; Ward and Duren, "Middle."

197 "the living primates of Africa": J. Kingdon, *The Kingdon Field Guide to African Mammals* (New York: Academic Press, 1997.)

199 "When I look at the postcranial bones": Interview with Shipman quoted in Shipman, "Muddle," p. 6.

10 Something to Chew On

201 "key life events correlate closely with both body size and brain size": See, e.g., R. D. Martin, *Human Brain Evolution in an Ecological Context,* 52nd James Arthur lecture, (New York: American Museum of Natural History, 1983); E. Armstrong, "Relative Brain Size and Metabolism in Mammals," *Science* 220 (1983): 1302–1304.

201 "infancy ends and the subsequent juvenile phase": A. Schultz, "Age Classes in Primates and Their Modification in Man," In J. M. Tanner, ed., *Human Growth* (New York: Pergamon, 1960), 1–20.

203 "adolescence, which is a peculiarly human invention": B. Bogin, "The Evolution of Human Childhood," *Bioscience* 40 (1990): 16–25.

204 "a large database on the dental development": B. H. Smith, "Dental Development as a Measure of Life History in Primates," *Evolution* 43 (1989): 683–688; B. H. Smith, "Dental Development and the Evolu-

tion of Life History," *American Journal of Physical Anthropology* 86 (1991): 157–174; B. H. Smith, T. L. Crummet, and K. L. Brandt, "Age of Eruption of Primate Teeth; A Compendium for Aging Individuals and Comparing Life Histories," *Yearbook of Physical Anthropology* 37 (1994): 1777–231.

205 "a shift in life-history strategies": J. Kelley, "Paleobiological and Phylogenetic Significance of Life History in Miocene Hominoids," in D. R. Begun, C. V. Ward, and M. D. Rose, ed., *Function, Phylogeny, and Fossils: Miocene Hominoid Evolution and Adaptations* (New York: Plenum, 1997), 173–208.

206 "Teeth carry their past": M. C. Dean, personal communication to Pat Shipman, 2003.

208 "the mineralization of enamel and dentine": See the reviews by T. Bromage, "Enamel Incremental Periodicity in the Pig-Tailed Macaque; A Polychrome Fluorescent Labeling Study of Dental Hard Tissues," *American Journal of Physical Anthropology* 86 (1991): 205–214; M. C. Dean, "Growth Layers and Incremental Markings in Hard Tissues: A Review of the Literature and Some Preliminary Observations about Enamel Structure in *Paranthropus boisei*," *Journal of Human Evolution* 16 (1987): 157–172; M. C. Dean, "The Developing Dentition and Tooth Structure in Primates," *Folia Primatologica* 53 (1989): 160–177; M. C. Dean, "Progress in Understanding Hominoid Dental Development," *Journal of Anatomy* 197 (2000): 77–101; M. Okada, "Hard Tissues of Animal Body: Highly Interesting Details of Nippon Studies in Periodic Patterns of Hard Tissues Are Described," *Shanghai Evening Post, Medical Edition,* Sept. 1943, 15–31; H. Shinoda, "Faithful Records of Biological Rhythms in Hard Tissues," *Chemistry Today* 162 (1984): 43–50 (in Japanese).

209 "long-period striae may represent": Dean, "Progress."

209 "a consistent, average periodicity": T. G. Bromage and M. C. Dean, "Re-Evaluation of the Age at Death of Plio-Pleistocene Fossil Hominids," *Nature* 317 (1985): 525–528; Dean, "Growth Layers."

210 "schoolsfull of children": M. C. Dean, interview with Alan Walker, April 2003.

211 "apes and humans spent the same number of days": M. C. Dean and B. A. Wood, "Developing Pongid Dentition and Its Use for Ageing

Individual Crania in Comparative Cross-Sectional Growth Studies," *Folia Primatologica* 36 (1981): 111–127.

212 "Once you sit down at the microscope": Don Reid, interview with Alan Walker, April 2003.

213 "a peculiar beauty to the histological structure": M. C. Dean, interview with Alan Walker, April 2003.

214 "the crown of the first molar"; "not incompatible with": A. D. Beynon, M. C. Dean, M. G. Leakey, D. J. Reid, and A. Walker, "Comparative Dental Development and Microstructure of *Proconsul* Teeth from Rusinga Island, Kenya," *Journal of Human Evolution* 35 (1998): 202.

214 "too little root to make the tooth functional": M. C. Dean, personal communication to Pat Shipman, 2003; Jay Kelley, personal communication to Pat Shipman, 2003.

215 "Wendy Dirks of Emory University": W. Dirks, "Histological Reconstruction of Dental Development and Age of Death in a Juvenile Gibbon *(Hylobates lar),*" *Journal of Human Evolution* 35 (1998): 411–425; W. Dirks, D. J. Reid, C. J. Jolly, J. E. Phillips-Conroy, and F. E. Brett, "Out of the Mouths of Baboons: Stress, Life History, and Dental Development in the Awash National Park Hybrid Zone," *American Journal of Physical Anthropology* 118 (2002): 239–252; W. Dirks, "The Effect of Diet on Dental Development in Four Species of Catarrhine Primates," *American Journal of Primatology* 61 (2003): 29–40.

11 More on Teeth

222 "Holly Smith developed another technique": B. H. Smith, "The Physiological Age of KNM-WT 15000," in A. Walker and R. E. Leakey, ed., *The Nariokotome* Homo erectus *Skeleton* (Cambridge: Harvard University Press, 1993), 195–220.

224 "Chimpanzees, humans, and *Proconsul*": A. D. Beynon, M. C. Dean, M. G. Leakey, D. J. Reid, and A. Walker, "Comparative Dental Development and Microstructure of *Proconsul* Teeth from Rusinga Island, Kenya," *Journal of Human Evolution* 35 (1998): 194.

225 "Though the enamel grew fast": Beynon et al., "Comparative," 197.

225 "The team estimated M1 eruption": J. Kelley, "Age at First Molar Emergence in *Afropithecus turkanensis,*" *American Journal of Physical Anthropology* Supplement 28 (1999): 167; J. Kelley, "Life History Evo-

lution in Miocene and Extant Apes," in N. Minugh-Purvis and K. McNamara, ed., *Human Evolution through Developmental Change* (Baltimore: The Johns Hopkins University Press, 2000), 223–248; T. M. Smith, L. B. Martin, and M. G. Leakey, "Enamel Thickness, Microstucture, and Development in *Afropithecus turkanensis*," *Journal of Human Evolution* 44, no. 3 (2003): 283–306; J. Kelley and T. M. Smith, "Age at First Molar Emergence in Early Miocene *Afropithecus turkanensis* and Life-History Evolution in the Hominoidea," *Journal of Human Evolution* 44, no. 3 (2003): 307–329.

227 "sectioning two isolated molars of *Victoriapithecus*": M. C. Dean and M. G. Leakey, "Enamel and Dentine Development and the Life History Profile of *Victoriapithecus macinnesi* from Maboko Island, Kenya," *Journal of Human Evolution,* forthcoming.

228 "as Benefit has hypothesized": B. R. Benefit, "The Permanent Dentition and Phylogenetic Position of *Victoriapithecus* from Maboko Island, Kenya," *Journal of Human Evolution* 25 (1993): 83–172

228 "David Begun and others": C.-B. Stewart and T. Disotell, "Primate Evolution—In and Out of Africa," *Current Biology* 8 (1998): 582–588; P. J. Heizmann and D. R. Begun, "The Oldest Eurasian Hominoid," *Journal of Human Evolution* 41 (2001): 463–481.

229 "we can turn to molecular evidence": R. L. Stauffer, A. Walker, O. Ryder, M. Lyons-Weiler, and S. B. Hedges, "Human and Ape Molecular Clocks and Contraints on Paleontological Hypotheses," *Journal of Heredity* 92, no. 6 (2001): 469–474.

230 "two other calibration points": S. Kumar and S. B. Hedges, "A Molecular Timescale for Vertebrate Evolution," *Nature* 392 (1998): 917–920.

232 "has speeded up over time": D. Penny, R. P. Murray-McIntosh, and M. D. Hendy, "Estimating Times of Divergence with a Change of Rate: The Orangutan/African Ape Divergence," *Molecular and Biological Evolution* 15 (1998): 608–610.

232 "the same branching sequence . . . similar dates for the divergences": Stauffer et al., "Human and Ape."

234 "it may be too derived or specialized": David Begun, personal communication to Pat Shipman, Jan. 2, 2004.

235 "*Lufengpithecus* might have been ancestral": Y. Chimanee, D. Jolly, M.

Benammi, P. Jafforeau, D. Duzer, I. Moussa, and J. J. Jaeger, "A Middle Miocene Hominoid from Thailand and Orangutan Origins," *Nature* 422 (2003): 60–65.

235 "the most likely orangutan ancestor yet found": Y. Chaimanee, V. Suteethorn, P. Jintasakul, C. Vidthayanon, B. Marandat, and J. J. Jaeger, "A New Orangutan Relative from the Late Miocene of Thailand," *Nature* 427 (2004): 439–441.

236 "*Sivapithecus* limb bones": D. Pilbeam, M. D. Rose, J. C. Barry, and S. M. Ibrahim Shah, "New *Sivapithecus* Humeri from Pakistan and the Relationship of *Sivapithecus* and *Pongo*," *Nature* 348 (1990): 237–239.

236 "another ambiguous case involving *Morotopithecus*," D. Gebo, L. MacClatchy, R. Kityo, A. Deino., J. Kingston, and D. Pilbeam, "A Hominoid Genus from the Early Miocene of Uganda," *Science* 276 (1997):401–404.

12 Listening to the Past

241 "The earlier anatomists": See, for example, A. C. Brown, "On the Sense of Rotation and the Anatomy and Physiology of the Semicircular Canals of the Internal Ear," *Journal of Anatomy and Physiology* 8 (1974): 327–333; A. A. Gray, *The Labyrinth of Animals*, vols. 1 and 2 (London: Churchill, 1907–1908); V. Tanturri, "Zur Anatomie und Physiologie des Labyrinthes der Voegel," *Maschinenschriftlich Ohrenheilkunde Laryngo-Rhinologie* 67 (1933): 1–27; H. J. Watt, "Dimensions of the Labyrinth Correlated," *Proceedings of the Royal Society of London* B 96 (1924): 334–338.

243 "specific features of the semicircular canals": A recent review can be found in F. Spoor, "The Semicircular Canal System and Locomotor Behavior with Special Reference to Hominin Evolution," *Courier Forschungsinstitut Senckenberg* (Bad Homburg: Forschungsinstitut Senckenberg, 1999), 1–13.

Index

Page references to illustrations are in italics